THE INTENT DRIVEN ORGANIZATION

MOVING FROM PROCESS AND ROLES TO POLICY AND OUTCOMES

ANDY FORBES

Roadmap Press
Hillsboro, Virginia, United States

ISBN: 978-1-969489-18-1 (paperback)
ISBN: 978-1-969489-17-4 (ebook)

Cover design by: Roadmap Press

To my managers at multiple companies who sparked my interest in management theory. And to the current leaders in Artificial Intelligence, whose innovations have made the reality discussed in these pages possible.

Andy Forbes

Be not the first by whom the new are tried, Nor yet the last to lay the old aside.

— ALEXANDER POPE

CONTENTS

PREFACE

Something quietly significant is happening inside many organizations. This transformation isn't announced with fanfare or breaking news. Instead it's woven into daily routines, revealing itself in small efficiencies and shifts rather than big headline moments. Advanced technologies, especially in artificial intelligence (AI), are beginning to take on tasks that once fell solely to people. Yet, amid all the talk about tools and innovations, there is limited discussion of what these changes mean for the design and management of organizations themselves. This book is an attempt to have that discussion. It focuses on the structural shifts and new forms of responsibility emerging in the workplace, rather than on the technology per se.

At its heart, this book is about how responsibility, control, and execution are changing inside modern organizations. These changes are driven by AI, but this is not a book about AI tools, platforms, or implementation techniques. It does not dwell on the mechanics of technology. Plenty of books and articles detail the latest AI tools; what is missing is guidance on how organizational life itself should adapt. This book aims to fill that gap by looking at how work and management must evolve internally. Indeed, many early AI projects in companies falter not for lack of technology, but because they try to overlay new tools onto unchanged structures. The workflows, approval chains, and definitions of roles stay the same, and so the promised improvements never fully materialize. Recognizing that pattern, this book centers on the adjustments in organization and responsibility that make new capabilities actually work. Instead of exploring algorithms, it examines how the presence of new capabilities alters the fundamental structure of how work gets done, how teams coordinate, and how decisions are made. The focus is on organizational structure rather than technical detail, and on shifts in how we assign responsibility and maintain control rather than on automation for its own sake. This emphasis on structure is intentional: even the most powerful new tool will falter if an organization's roles and rules aren't set up to support it. By contrast, a sound structure can amplify the value of new technology. In practice, it's the policies, processes, and clarity of accountability inside an organization that determine whether innovations actually improve outcomes. Thus, the pages ahead concentrate on those internal factors that make the difference between just having new capabilities and truly harnessing them.

This perspective means looking at every function where outcomes matter. Whether in finance or legal, operations or human resources, the same structural forces are at play. The ideas here are not confined to any one department or industry. They speak to a broad organizational shift that affects anyone accountable for results. By design, the discussion is rooted in everyday organizational life, not in theory or speculation, so that leaders and practitioners across disciplines can recognize the situations described. It does not matter if you work in a bank, a hospital, a manufacturer, a government agency, or a technology firm the underlying structural questions are similar wherever people work together toward important goals. The specifics may vary by field - a bank might look at how to automate parts of its compliance checks, a manufacturing firm might rethink its quality control processes, a hospital might introduce AI assistants for routine patient queries - but the structural questions of responsibility and oversight remain remarkably alike. This is a book for those who manage real processes and deliver real outcomes, who sense that familiar ways of working are encountering new kinds of friction. Many current frustrations within companies and institutions stem from a model that still works, but not as cleanly as it once did. That subtle loss of fit is often a sign that something fundamental is shifting. More and more, teams find themselves saying that a process feels slower than it should be, or that rules which once ensured quality now sometimes hinder speed or innovation. Such everyday frustrations are usually the first signals that the environment has evolved beyond the current organizational design.

Not long ago, it was assumed that no complex operation could run without human oversight somewhere in the loop. But that assumption is now eroding. Beneath these emerging changes lies a single assumption that this book adopts as a starting point. This book proceeds from one explicit premise: certain forms of execution can now, or very shortly, be reliably delegated to machines or software, under well-defined rules and constraints (in other words, explicit policies and guardrails that tell the system what it may and may not do), without continuous human oversight. In plain terms, some tasks can be performed consistently and safely by automated systems following explicit policies, freeing people from having to supervise every detail. Automation itself is not new in organizations, but what is new is the level of autonomy: these systems can carry out their functions without a person watching over each moment. We do not attempt to prove this premise or explain the technology behind it; we simply state it as the context for what follows. If it were to prove untrue, the structural arguments in this book would not hold. Because we take it to be directionally true, our focus is on the organizational implications rather than the mechanics. This is the first and last time we will articulate this premise so directly; beyond this point, it simply serves as a background condition for the structural exploration. To be sure, not every job or process can be handed off to machines, far from it. But enough routine, rule-based activities now can be automated that the overall design of how work gets managed must adapt accordingly. Even without fanfare, this change in capability is beginning to have an impact. Its effects can be felt in subtle ways across organizations, sometimes even before leaders have fully recognized them. A software bot might silently approve an expense request that meets all policy criteria or flag an anomaly in data for review, tasks that once would have waited for human attention. This premise may feel like a leap for those of us trained to always keep a watchful eye on operations, but it is not presented here as hype. It is simply the condition under which we will explore the organizational consequences. From here onward, we treat it as a given.

This book is intended for people across all organizational functions who carry real responsibility for outcomes. These readers might be executives or managers, but they are also senior specialists or team leads. What matters is not rank or technical background, but the fact that they are accountable, personally and directly, for decisions, risks, and results. A leader in finance managing critical budgets, a director in operations overseeing key processes, a project manager in IT delivering complex initiatives, or a department head in HR guiding workforce changes - they all share the common burden of making sure that what needs to happen does happen. They know the pressure of meeting targets, complying with regulations, protecting their teams, and delivering on promises. They are the ones on the hook when goals are missed or when risks materialize, and that reality informs their perspective. The content here speaks to anyone who knows the weight of being answerable for performance.

Because the audience is defined by accountability rather than by job title, the perspective of the book remains practical and grounded. It treats the reader as a peer, as someone who understands the constraints and complexities of real organizations. There is no flattery and no hand-holding, only a recognition of shared challenges. Seasoned readers in this group have seen management trends come and go; their skepticism of grand promises is well-earned. This book acknowledges that history and aims to offer something more solid than the latest fad. Those who see themselves in this description will likely recognize many of the situations and tensions explored in the chapters ahead. The problems discussed in these pages are common to responsible leaders across industries, and the analysis is intended to resonate with their experience. If you have felt a sense that your organization's old way of working is a bit out of tune with new realities, you are far from alone because many leaders quietly share that feeling.

Readers outside this profile may find the approach here unfamiliar. This is not a book for those seeking a crash course in AI technology, terminology, or a quick-fix checklist for modernization. It will likely frustrate someone whose focus is purely technical or who expects instant solutions. If you come looking for coding advice or a step-by-step playbook, you will not find it in these pages. You will not find a generic five-step roadmap to become "AI-driven". Every organization's path is different, and simplistic formulas would only gloss over hard realities.

This book does not encourage reckless change. It does not urge anyone to dismantle effective processes or abandon the safeguards that manage risk. The approach throughout is measured. There is no "change now or perish" slogan here, just as there is no naïve optimism that technology automatically fixes everything. Restraint and clarity are intentional choices. Where others might push urgency and disruption, this discussion leans into understanding and careful evaluation. By setting these boundaries, the aim is to reduce anxiety and keep the focus on fundamentals. Nothing important is omitted; on the contrary, the omissions are deliberate, made to avoid noise and distraction. If a quick formula or dramatic call to action is missing, it is by design. That absence reflects a respect for the complexity of organizational change and a belief that clear thinking must precede bold moves. It also recognizes that organizations differ greatly; there is no single formula that fits everyone. Instead of pretending there is one right answer, the book offers frameworks to help each reader think through what might be right for their situation. What might seem like gaps, such as the lack of a simple formula to follow, are in fact deliberate. They reflect a respect for the complexity of organizational change and a belief that understanding must come before decisive action.

Modern organizations did not take their shape by accident. The structures we know - layers of management, defined departments, sequential processes, approval checkpoints, and oversight mechanisms - were all designed for a world where execution depended entirely on human beings and on human effort, judgment, and attention. Organizations have been built, first and foremost, to compensate for human limitations. People are fallible: they get tired, they vary in judgment, they cannot pay attention to everything, and they make mistakes especially when coordinating at scale. We all know that errors tend to multiply in the late hours of a long shift or when tasks outstrip a person's bandwidth. Recognizing those limits, organizations built in safeguards. So we put in place checks and balances. A budget proposal might need a second approval from someone else in Finance; a new contract gets a legal review before it is signed; a production change follows a standard procedure so that safety and quality are not left to chance. In short, we added structure to achieve reliable results despite the natural variability of human performance. This design was rational and successful. It allowed complex organizations to function with a high degree of predictability and safety, even as they grew larger and more complicated. It gave managers, employees, and customers alike confidence that the enterprise could deliver on its promises consistently, without having to rely on any single individual's heroics.

Now, however, the environment is changing in ways that strain this once-effective model. Some of the human constraints that shaped our systems are beginning to ease, and new pressures are emerging that the old structure was not built to address. The very mechanisms that were essential to maintaining control and consistency can start to feel cumbersome or slow under these new conditions. In response, the mechanisms organizations use to stay in control are gradually shifting. ("Control" in this context isn't about power for its own sake, but about the means by which leaders make sure that outcomes meet expectations.) Broadly speaking, the center of gravity is moving from rigid process to flexible policy, from static roles to fluid decision-making, and from constant supervision to guided intent.

Instead of relying so heavily on step-by-step procedures and multiple layers of approval, organizations can increasingly set clear policies and guidelines that intelligent systems and well-prepared people follow. Instead of carving up work strictly by narrow job descriptions, they can pay more attention to the decisions that need to be made and ensure those decisions are made by the right people (or even machines) at the right time. Instead of managers overseeing each small task, leaders can articulate a clear intent and trust teams and technology to carry it out within well-defined guardrails. These shifts may start subtly, but over time they amount to a fundamental rebalancing of how coordination and control happen. By making these changes, organizations can become more adaptive: broad policies guide action without choking initiative, decision-focused roles respond faster to new needs, and clear intent allows execution to proceed without constant check-ins.

None of this implies that the old model was flawed or that human judgment is becoming obsolete. On the contrary, the traditional structure exists for good reasons, and many of its elements remain vital. The goal of structure e.g. ensuring work is controlled and outcomes are reliable, remains the same, even as the methods change. This book takes that stance throughout: it respects why the current model exists and those who have been successful within it. Nothing here is about faulting the past or casting aside the wisdom that got us this far. Accountability does not disappear; if anything, it becomes even more important to define who is responsible for setting policies, making key decisions, and establishing intent. Likewise, human expertise remains indispensable. This evolution is not about phasing out human judgment or labor, but about using it more wisely. People are still needed to design those guiding policies, to interpret the results that automated systems produce, and to manage the exceptions and complexities that no algorithm can foresee. But the balance is changing. What was once the only optimal way to organize work is no longer the only way and in some cases is not the best way under new circumstances. The core argument of this book is that these structural shifts are real and increasingly unavoidable. They are driven by practical pressures and emerging opportunities, not by any fashionable theory.

This would not be the first time that technology forced a rethink of organizational design. One can point to earlier eras: for example, the introduction of assembly line production in factories, which reorganized work and required new supervisory methods; or the spread of computers in offices, which flattened some layers of communication and changed how information flowed. In each of those cases, organizations that adapted their structures to the new reality thrived, while those that clung too long to old models had to, sometimes painfully, catch up later. The emergence of autonomous execution capabilities is another inflection point of similar magnitude, even if its effects are unfolding gradually.

This is not a call for abrupt revolution. Change in organizational design can be thoughtful and incremental. By understanding how and why things are shifting, leaders can adapt without panic or haste. The chapters that follow examine this evolution step by step, making the case that while the shift in how we execute work is profound, it can be approached with a calm, informed mindset. The move from process to policy, from roles to decisions, and from supervision to intent can be an evolution that can be navigated responsibly, with respect for what came before.

The changes outlined here are not speculation about a distant future. They are already quietly underway. In some organizations, small pilot projects have begun to let automated systems handle tasks that once required multiple human checkpoints. An accounts department might program a system to automatically approve routine expenses under a certain dollar amount. What used to sit in a manager's inbox for a day now gets cleared in seconds by a script following company policy. The safeguard still exists, but it's enforced by software instantly rather than by a manager's signature a day later. Elsewhere in the organization, a few teams are experimenting with more flexible, principle-based workflows in place of rigid step-by-step procedures. These developments are often piecemeal and not widely advertised. Minor inefficiencies or delays might be written off as routine bureaucracy rather than recognized as symptoms of a deeper change. A manager might see an increase in exceptions to a process and assume it's a temporary blip, not realizing it signals a mismatch between old procedures and new possibilities. There is no grand announcement that the structure of work has changed; most of the old mechanisms remain in place, carrying on as they always have. Because there is no obvious crisis or mandate to change, it is easy for these early signals to be overlooked, or for leaders to decide they can simply wait and see.

For many, "if it isn't broken, don't fix it" is a reasonable policy. But by the time something is clearly broken, fixing it can be far more difficult. Paradoxically, this lack of urgency makes now the ideal time to pay attention. Structural shifts in how organizations operate tend to unfold gradually, until one day adaptation becomes unavoidable. By the time a change is obvious to everyone, the adjustments required may be more difficult and rushed. To illustrate, imagine two companies confronted with the same new capability: one proactively adapts its processes and policies in anticipation, while the other waits until market forces demand change. The first company can integrate the new methods gradually, training people and refining rules with less pressure. The second, having delayed, must scramble in crisis mode and implement changes hurriedly, with greater risk of disruption. This simplified contrast underscores the advantage of engaging with structural shifts before they reach a boiling point. Today, we have the benefit of time and relative stability - the conditions under which reflection is possible. This book is timely because it seeks to name and examine these patterns before they turn into pressures that demand a reaction. Rather than sounding an alarm or insisting on immediate disruption, the aim is to give readers a clear view of what is happening so they can consider their course deliberately. In an atmosphere free of panic, it is easier to think through which traditions to hold on to and which new approaches might be sensible. Responsible leaders often choose to engage with important changes early, not because they are forced to, but because they recognize subtle signs and prefer to prepare. Engaging with a structural change before it becomes a crisis is part of prudent stewardship. In that spirit, taking the opportunity now to understand the evolving model of execution is a way to lead, rather than to chase events after the fact. Making changes before a crisis isn't just prudent, it's a defining quality of effective leadership.

This book is designed to be read as a progression of ideas, building understanding step by step. It begins by laying a groundwork. The opening chapters examine the current organizational model on its own terms, showing how it works and why it made sense for its time. After establishing that foundation, the focus shifts to the cracks starting to appear under new conditions. The middle sections of the book diagnose where traditional structures are under strain and explore why those stresses are emerging, not to find fault with anyone, but to understand the systemic causes. Finally, with the problem context defined, the later chapters consider how the pillars of responsibility and control might evolve. These chapters introduce adjustments in perspective such as moving from detailed process to guiding policy, from fixed roles to flexible decision rights, and from active supervision to defined intent. Throughout this journey, the emphasis stays on structural principles rather than any specific technologies or management fads. Each part builds directly on the one before, so that understanding deepens gradually and logically. After exploring these shifts, the book also takes time to consider the risks on both sides of change both highlighting the pitfalls of moving too aggressively and the dangers of standing still too long. The closing chapter then offers a balanced perspective on how to approach the transition, emphasizing continuity even amid change. In effect, the book leads the reader from understanding why things are the way they are, to seeing why change is brewing, to reimagining key elements of the model, and finally to considering when and how to navigate the transition. Each stage builds on the prior, making the journey logical and complete.

Because of this deliberate structure, the book is meant to be read in order from start to finish. Each chapter builds on concepts from previous chapters, and ideas introduced early on are not re-explained later. Key terms and concepts appear consistently throughout; once defined, they will be referred back to in the same language rather than rephrased or repeated at length. This means an attentive reading of the early chapters will pay dividends in understanding the later ones. Skipping ahead or browsing out of sequence would risk missing pieces of the argument and undermining its coherence. This is not a book to dip into for isolated tips or snippets of advice. It's not a lightweight manifesto to skim on a short flight. Rather, it's more like a guided analysis session that rewards a careful, thoughtful reading. Instead of quick answers, it offers a building argument. The pace is intentional, calm and methodical, so that by the end the reader can see the whole picture that has been assembled. Taking the chapters in sequence and allowing each idea to unfold will provide a far clearer understanding than trying to cherry-pick topics. As you move through each chapter, consider your own past projects or decisions. You will likely recognize moments when a well-designed process saved the day, or when a cumbersome approval chain caused a missed opportunity. Connecting the concepts to real incidents in your experience will make them more concrete and actionable.

By the final page, a reader should have a clearer view of their organization and how it might adapt in light of these shifts. The payoff for following the argument through is a refined mental model of how complex operations fit together, why they were designed as they were, and how they could be reimagined. With that comes a more precise language for thinking and talking about change. Concepts that might otherwise remain vague - "automation," "AI-driven processes," and the like - can be discussed in concrete terms of policies, decisions, roles, and accountability. The goal is not to hand over a formula or a simplistic solution, but to equip readers with a deeper understanding of the dynamics at play. It will not erase uncertainty, but it will replace some of the guesswork with insight. With greater insight, you may also find a new patience for the complexities in your organization. What once seemed like pointless bureaucracy can be seen in context, as a necessary adaptation of its time, making it easier to discuss improvements without frustration or blame.

Practically speaking, this clarity will help in making judicious decisions about if, when, and how to evolve an operation. It provides a basis for deciding which new capabilities to embrace and which to approach cautiously. Readers should be better prepared to judge where greater flexibility can be introduced safely and where firm structure should be maintained, which responsibilities might be delegated to automated systems and which must remain firmly in human hands. In short, the outcome of reading is not a checklist of actions, but a stronger grasp of organizational mechanics and the trade-offs that come with changing them. It means having a more nuanced sense of timing and risk and recognizing when a bold change is worth the gamble and when restraint is the wiser course. For example, this book will not tell a manufacturing executive or a hospital administrator exactly which processes to automate next year; instead, it will help them weigh the decision for their specific context by illuminating the structural factors at play. That clarity is meant to foster sound judgment. A leader who has thought through these chapters will be able to approach structural changes with eyes open and a steady hand, rather than being driven by hype or anxiety. The ultimate aim is to enable more grounded and responsible decision-making in an environment that is evolving in ways we can understand and manage.

In practical terms, the reader may find themselves approaching their next reorganization effort or technology proposal with a new lens. Instead of simply asking what a new tool can do, they will ask how work is structured and where accountability will lie once that tool is in place. For example, if someone suggests automating a routine process, a reader of this book will be ready to probe deeper: Who will monitor the outcomes? Do our policies explicitly cover this scenario? Are we transferring a decision from a person to a system, and if so, how will we maintain oversight? This kind of nuanced inquiry is exactly what the following chapters aim to encourage.

No matter how sophisticated our tools become, effective organizations will always depend on deliberate design and management. Technology can change what's possible, but it takes human insight to decide how to harness those possibilities. In short, tools alone don't run companies, people do, and it's the structure we create around those tools that makes all the difference. It's a fundamental truth that can be overshadowed by technology hype, but it remains crucial.

The first step in that understanding is to look squarely at the model we have relied on up to now. Before considering any new direction, we must first understand why our organizations look and operate the way they do and how human limitations shaped the structures we take for granted. In the pages ahead, we start at the beginning: by examining the familiar foundations of the current model and the logic behind them. Only on that firm ground can we properly consider what might come next.

1

THE ORGANIZATION WE BUILT FOR HUMAN LIMITS

Every large organization today, regardless of industry, shares familiar structural features: layers of management, specialized departments, formal approval processes, and clearly defined roles. These similarities are not coincidental or simply traditional. They exist because modern organizations have been deliberately shaped to compensate for a fundamental reality, the predictable limits of human beings in how we work. In effect, we designed our institutions around human constraints. Individual people, even the most skilled and diligent, have limited attention spans and can only handle so much complexity at once. We get tired, we make occasional mistakes, and no one person can be everywhere or know everything. As organizations grow in size and scope, those human limitations become more pronounced. To deliver reliable results at scale despite these constraints, companies introduced structures that act as shock absorbers and guide rails. The hierarchies, roles, handoffs, and rules we take for granted are compensating mechanisms: they make outcomes predictable even when individual behavior might not be.

We can see this logic in every function of the enterprise. In a finance department, for example, no single employee can approve a major expenditure on their own; large purchases require multiple sign-offs. This isn't because individuals can't be trusted at all, it's because a second pair of eyes reduces the chance of oversight or error. Likewise, legal teams institute multi-step reviews for important contracts, ensuring that different experts examine the language so nothing critical is missed. In operations, front-line staff follow escalation paths for significant issues: a factory supervisor is expected to hand off a serious safety problem to higher management rather than make an isolated judgment call. In HR, hiring decisions typically involve sequential interviews by different people, background checks, and committee discussions, rather than a single manager's whim. Across these very different domains, the pattern is the same. Organizational structure creates checks and balances that no individual, with their finite judgment and perspective, could achieve alone. Roles segment complex tasks into manageable pieces, handoffs ensure accountability at each stage, and reporting lines formalize who is responsible for what. The ultimate goal is to engineer consistency out of human variability, to make sure that the work gets done to a reliable standard, day after day, even though humans are inconsistent and fallible by nature.

What's striking is that these structural solutions appear in all kinds of organizations, not just in one industry or culture. A bank, a hospital, a tech company, and a government agency will all show a hierarchy of authority and a division of labor that look surprisingly alike. This convergence is a strong hint that the driving force is something universal. Indeed, hierarchical organization has endured for centuries and is arguably "the most common human organization system of all," found in everything from religious institutions to military units to modern corporations. The reason is straightforward: humans in large groups need structure to coordinate effectively. Without formal structure, once a group grows beyond a certain size, communication breaks down and chaos looms. Anthropologists have noted, for instance, that humans can maintain only a limited number of close, stable relationships, often cited around 150 people, before social cohesion requires formal rules and enforced norms. In other words, beyond a modest scale, you cannot rely on personal trust and ad-hoc interaction alone to run an organization; you need explicit roles and procedures. Our organizations grew large by adding structure to overcome natural cognitive and social limits. Hierarchy, departments, and standard operating procedures filled the gap between what individuals can manage informally and what the enterprise needs to accomplish at scale.

Seen in this light, the classic components of organizational design are not arbitrary bureaucracy; they are rational responses to human constraints. Dividing work into specialized roles recognizes that no individual can master every skill or attend to every detail. Limiting the span of control, the number of direct reports under one manager, acknowledges that there is a cognitive cap on how many people one can supervise effectively (a widely cited rule of thumb is around seven, give or take a couple). Requiring oversight and approval for critical tasks accepts the fact that human judgment varies and benefits from a second opinion. All of these practices were put in place to ensure that the output of the whole remains reliable, even if any one person might have a bad day or a blind spot. Modern organizations, in essence, built a kind of external scaffolding around our human limitations. The structure catches our mistakes, spreads out our workloads, and links us together in a stable configuration that can produce far more than a collection of uncoordinated individuals ever could.

It's important to recognize that these designs, which now feel so fundamental to "how work is done," arose to solve particular problems. They feel permanent, but they are contingent on the limitations they were meant to address. We often assume that a clear chain of command or a rigid approval chain is just inherently necessary. In truth, it became necessary because of how we as humans operate. What feels like bedrock in organizational life, the bureaucracy, the org charts, the rulebooks, is in fact a product of circumstance. It was built for a world where humans are the ones driving every process. This does not diminish its value; if anything, it highlights the brilliance of the solution. But it also introduces an intriguing thought: if the circumstances were to change, if the human limitations that defined the old problems were altered, then the structures we inherited might start to look less inevitable. In short, what we consider fundamental in our organizations is actually situational. The only way to appreciate that is to understand exactly why the current model exists. And that starts by examining how human limitations led directly to the structures we now have.

HUMAN INCONSISTENCY MADE STRUCTURE NECESSARY

A core premise of this chapter is that inconsistency, not individual incompetence or ill will, is what drove the need for so much structure. No matter how talented or well-intentioned people are, their judgments and performance will vary. Two employees handling the same task might do it in slightly different ways; one project manager might interpret a policy more liberally than another. One salesperson might take risks that another avoids. These variances are natural, they are a byproduct of human diversity and the simple fact that no two people think exactly alike. In a small team, such differences can often be managed informally. But as soon as you scale up the work or the stakes, variance turns into risk. Without coordination mechanisms, one client might get excellent service from a diligent representative while another gets mediocre attention due to an oversight. One factory shift might produce high-quality output, and the next shift, with a different crew, produces defects. Early industrialists and organizational leaders learned by hard experience that relying on individual heroics or personal style didn't work when you needed consistent, repeatable outcomes. So they engineered consistency through structure.

The key insight here is that this was not about mistrusting people or curbing freedom for its own sake. It was a practical solution to a practical problem: unbridled inconsistency leads to unpredictable results, and unpredictable results are unacceptable when you're responsible for outcomes at scale. For example, if every regional office of a company interpreted accounting rules differently, the consolidated financial statements would be a mess and likely inaccurate. Therefore, standard procedures and centralized policies emerged, not because the head office didn't "trust" local accountants, but because without a unifying structure the numbers simply wouldn't add up. Similarly, consider safety regulations on a factory floor. One worker might be less attentive to a safety checklist than another, perhaps not out of negligence but simple human variability or fatigue. Rather than hope that every single individual is always vigilant (an unrealistic expectation), companies put in oversight, a floor supervisor double-checking that protocols are followed, or an automated system that flags a missed step. The structure catches the inevitable slip-ups or differences in execution before they cause harm.

Over time, organizations found countless ways to buffer against human variability. Coordination mechanisms, things like cross-checks, peer review, and management approvals, became baked into the workflow. A core belief behind these mechanisms is that errors and lapses are not moral failings; they are statistical probabilities. Humans are fallible, and even the best people will occasionally forget, overlook, or misjudge something. High-reliability industries (like aviation and healthcare) learned to treat human error as a systemic issue rather than a personal one: they build systems of defense in depth, expecting that people will miss things and designing multiple layers to catch problems before they cascade. This "system approach" to human fallibility views mistakes as consequences of the conditions under which people work, not as evidence of individual weakness. In practical terms, that means creating an environment where the same task is done the same way, no matter who performs it or when. Structure, through rules, routines, and oversight, is what creates that environment.

It's worth emphasizing that these compensations were rational, necessary, and remarkably effective. None of this is to say that people should be more consistent or less prone to error; expecting flawless performance from humans is futile. Instead, leaders chose to work with human nature rather than against it. They built organizations that assume variability and absorb it. When you walk into a well-run large enterprise, you see a thousand moving parts operating in concert. That concert is possible only because behind the scenes, the variability of each part is being managed, the procurement officer's interpretation of vendor risk is guided by policy; the call center rep's response to an angry customer is guided by a script or training; the junior engineer's design change is checked by a senior engineer in a design review. These are coordination tools to ensure reliability. Crucially, none of this means the people involved are not skilled or responsible. It means their individual judgments are augmented by a larger system to reduce the chance of a serious lapse. Structure, in this sense, is an insurance policy against the unpredictable quirks of individual behavior.

By acknowledging human inconsistency without blaming the humans, organizations created an atmosphere where process and structure became seen as forms of diligence rather than oppression. Good managers understood that the goal was not to police workers but to protect outcomes. In fact, a well-designed structure frees individuals to perform without the constant fear of personal failure, because they know the system will catch mistakes and provide guidance. The veteran pilot trusts air traffic control and standardized checklists; the skilled surgeon relies on the surgical team's established protocol and time-out procedure to ensure nothing is missed. These structures do not diminish the expertise of the pilot or surgeon, they amplify it by adding reliability. In business contexts, the same principle applies: an experienced project lead might chafe at bureaucracy, but also knows that having a defined stage-gate process and oversight can prevent costly errors that no one would want on their watch.

All of this reinforces respect for the existing system. The people operating within these structures deserve credit for mastering them and making them work. It is much harder to deliver consistent results than to occasionally deliver brilliant results amid inconsistency. The professionals in any large organization who manage to meet their targets year after year are usually those who understand how to leverage the structures, how to coordinate, escalate, double-check, and follow the necessary routines. They are not robots; they are skilled operators who know that structure is part of the job. By seeing structure as the enabler of reliability rather than a burden, they carry out their responsibilities with a high degree of confidence that no single human mistake (including their own) will sink the ship. This mindset, that inconsistency is a problem to be managed, not a personal flaw to be eradicated, is what made modern organizational life possible at scale. It set the stage for specific structural inventions like hierarchy, formal roles, and systematic handoffs, which we turn to next.

Before moving on, consider one more point of tension: the cost of compensating for inconsistency is not fixed. The more complex and large an organization becomes, the more structure it tends to pile on to maintain consistency. Each added rule or checkpoint might be necessary, but each also introduces a bit more overhead. At a small scale, coordination is easy and process light; at a massive scale, coordination requires heavier machinery. We will later explore whether those costs, the cost of consistency, are now being felt more acutely in many organizations. But first, we need to examine the specific mechanisms by which structure was implemented. Chief among those are hierarchy, roles, and the deliberate segmentation of work, all of which were coherent responses to the challenges of human limits.

HIERARCHY, ROLES, AND HANDOFFS WERE RATIONAL RESPONSES

It is no accident that virtually every sizeable organization is hierarchical. Hierarchy is perhaps the quintessential organizational response to human limits on decision-making and oversight. The premise is simple: rather than trying to have every decision made by one person or by a collective free-for-all, organizations delegate decisions to different levels of authority. This distribution of decision load ensures that no single individual is overwhelmed and that decisions are made by those closest to the relevant information. In published writing, Herbert Simon, the Nobel-winning scholar of organizations, pointed out that hierarchy allows each level to focus on decisions within its scope and expertise, making the overall system more workable than if one brain attempted to optimize everything centrally. In practical terms, the CEO doesn't approve every office supply purchase, and the front-line employee doesn't set corporate strategy. By breaking the work of thinking and deciding into tiers, hierarchy acknowledges the cognitive limits of any one manager and uses specialization of attention as a solution.

Roles, similarly, narrow the field of responsibility for each person. A role comes with a defined set of duties and boundaries. This clarity is a feature, not a bug: it prevents individuals from overstretching and ensures that critical functions are tended to by someone explicitly charged with them. Historically, the division of labor into roles has been celebrated as a source of efficiency and scale. Adam Smith's famous example of the pin factory showed how breaking the production of a simple pin into many small tasks (drawing the wire, cutting it, sharpening the point, etc.) multiplied output dramatically. One worker alone, switching between all tasks, could hardly produce anything; ten workers in specialized roles could produce tens of thousands of pins per day. The lesson was that specialization overcomes individual limitations, in skill, in time, in focus, by allowing each person to concentrate on a manageable aspect of the work. In a modern office, a payroll specialist isn't expected to also design marketing campaigns, and an IT security engineer isn't also negotiating vendor contracts. By narrowing responsibilities, each person can develop depth in their role, and the organization as a whole gains reliability and expertise in every necessary area.

Handoffs are the connectors between those specialized roles. Whenever work passes from one role to another, say, from a product designer to a manufacturing engineer, or from a sales representative to a fulfillment team, the handoff creates a clear accountability boundary. Each party is responsible for their segment, and the handoff itself is often governed by some check to ensure nothing critical falls through the cracks. One way to think about handoffs is as controlled transfers of responsibility. They mark the point where one person's human limits (be it time, expertise, or authority) are reached and another person's capabilities take over. If structured well, handoffs add resilience: if five people sequentially handle a process, a mistake by one can be caught by the next. The classic assembly line is an illustration, each station checks or relies on the quality of the previous station's work. In knowledge work, a report might be drafted by an analyst, then reviewed by a team lead, then approved by an executive. At each handoff, there's an opportunity for correction and alignment. This slows things down a bit, but it raises confidence that the end result meets the intended standard.

Hierarchy, roles, and handoffs work in concert to make large organizations governable. They address different facets of the human limitation problem: hierarchy deals with span of control and cognitive overload, roles deal with depth of skill and clarity of duty, and handoffs deal with sequential workload and error-checking. Together, these elements allowed enterprises to grow vastly larger and more complex than any flat, all-to-all communication network of people ever could. Imagine a company of 50,000 employees with no hierarchy, where anyone could make decisions about anything, it would be utter chaos. Instead, we have pyramids of authority for clarity. And within those pyramids, we have myriad distinct roles so that expertise can flourish and not everyone has to worry about everything. Work flows along defined channels from one role to the next, rather like an organism's circulatory system, ensuring each piece of work gets the right attention at the right stage.

Crucially, this organizational form, the one most of us inhabit daily, was not accidental, outdated, or poorly conceived. On the contrary, it was a coherent solution to very real constraints. Consider the alternative: if we had unlimited cognitive capacity and flawless communication skills, perhaps we wouldn't need hierarchy or formal structure; everyone could just organically coordinate. But in the real world, that's not possible. Hierarchy imposes an order that counters entropy. Defined roles counteract the impossibility of individual omniscience. Structured handoffs counteract the risk of individual error. The fact that similar structures arose in corporations, government bureaucracies, armies, and even large nonprofits underscores that these were rational responses to universal human limits. Max Weber, writing in the early 20th century, famously described bureaucracy (with its hierarchy, rules, and specialization) as the embodiment of rational organization, efficient, predictable, and capable of handling the "large amounts of information" and "complex activities" that modern life demands. While the word "bureaucracy" often has negative connotations, Weber's point was that a rule-bound hierarchy is technically superior to ad-hoc management for achieving consistent results at scale. It may feel impersonal, but it was born from necessity, not whim.

Of course, structure can have a downside: it can become rigid, and sometimes people feel boxed in by narrow roles or frustrated by layers of approval. Those are real issues, but it's important to remember that the coherence of this design is also its strength. By being coherent, by fitting all the pieces together in a logical way, the system as a whole maintained order and clarity in environments that would otherwise tip into confusion. Everyone knows who their boss is, what their job is, and where their work goes next. That clarity has enormous value; it produces alignment and reduces friction in daily operations. However, coherence in design can turn into rigidity if the assumptions underlying the design change. In other words, the very features that made our organizations so sensible and effective can become liabilities if the context shifts. We will not jump to conclusions yet about what might have changed, but it's a thought to hold: a structure optimized for one set of limitations might need re-examining under a new set. For now, what matters is that hierarchy, roles, and hand-offs achieved what they were intended to. They made the unwieldy task of coordinating thousands of humans manageable. They created governability out of potential chaos. And they did so through a logic that any responsible leader can respect: divide the work, delegate authority within clear bounds, and build in oversight as work flows upward.

OVERSIGHT BECAME THE EXPRESSION OF RESPONSIBILITY

One of the most visible features of traditional organization life is oversight. Managers review and approve the work of their subordinates. Executives demand reports and status updates. Auditors and quality assurance staff comb through processes to ensure compliance. This pervasive oversight is not just bureaucracy for bureaucracy's sake, it became the very definition of what it means to be responsible in a human-driven operation. In a company built on human labor and human decision-making, being accountable inevitably came to mean being the person who checks and corrects the work of others. The logic is intuitive: if humans are the "execution layer" of the business, then the role of a supervisor or an accountable leader is to watch that execution, guide it, and intervene when something goes wrong.

Historically, as organizations grew, senior leaders could not possibly do all the work themselves, their span of attention and time was limited, so they had to rely on others to execute tasks. But relying on others didn't mean ignoring what those others did; it meant reviewing and controlling it. This is how responsibility got operationalized as oversight. A chief financial officer is accountable for the integrity of the financial statements, but of course cannot personally prepare every entry, so they institute controls, sign-offs, and audits, effectively overseeing an entire accounting process. A VP of Sales is accountable for revenue targets, but cannot be on every sales call, so they review pipeline reports, listen in on some calls, and set policies for discounts and deal approvals. These oversight activities are how leaders ensure that the work under their charge meets the necessary standard. To be in charge, in the traditional model, is to be the ultimate failsafe, the person who will catch the error, make the tough call, or enforce the rule if needed.

Over time, this notion of oversight as responsibility became deeply embedded in organizational culture. Climbing the hierarchy meant accumulating oversight duties. A promotion to manager meant you now approve your team's output; a promotion to director meant you now oversee managers and set guidelines for them. Even in boards of directors, the top of the governance pyramid, much of the role is oversight of the executives (ensuring the CEO and team are doing the right things and following the right policies). This isn't about distrust; it's about having a second layer of consciousness on important work. We learned, sometimes painfully, that leaving critical decisions or processes entirely in one person's hands courts disaster if that person slips. The banking industry, for instance, enshrined the "four-eyes principle", requiring that key transactions or decisions be approved by at least two people, specifically to prevent errors or fraud that a single person might miss. In effect, formal oversight is a way of saying, "This outcome matters too much to depend on a single human's judgment or integrity." Therefore, multiple humans must be involved, with one watching over the other. It's an elegant use of redundancy: two imperfect actors can, together, produce a more perfect result.

For anyone who carries formal accountability in an organization, this dynamic is likely very familiar. If you are responsible for an outcome, whether it's a project delivery, a compliance program, or a customer experience, you instinctively create ways to monitor and verify the work that feeds into that outcome. You might set up regular review meetings, insist on being consulted for exceptions, or implement dashboards to track performance. Far from being meddling, these acts are usually seen (by the responsible person, at least) as fulfilling their duty. After all, if something goes wrong, it is the accountable person who answers for it. It's not surprising, then, that they feel the need to keep their finger on the pulse. Over decades, organizations routinized this: they created standard oversight checkpoints such as weekly status reviews, approval chains, performance appraisals, audit committees, etc. Being a leader became synonymous with diligently monitoring those you lead.

It's worth noting that this model was a structural necessity given the human-centric execution of work. If the work is being done by people, who, as we established, are variable and fallible, then oversight is your quality control mechanism. It's not just a cultural preference that managers should hover; it's a logical requirement to achieve consistency and correctness. In many industries, oversight is literally written into laws and regulations (think of pharmaceutical manufacturing requiring documented review of every batch, or financial trading requiring supervisory approval of certain transactions). These rules emerged because without oversight, the risk of individual error or misconduct was simply too high. So oversight can be seen as the organizational embodiment of caution. It is how an institution says: "We will trust but verify. We will delegate, but we won't abdicate." The person delegating still bears the responsibility, so they maintain a line of sight into what's happening below them, ready to step in if needed.

Having said all that, it's important to step back and reflect on whether responsibility must always manifest as oversight. We generally assumed it must, because for the last century or more, the only agents executing work were people. If that's the case, then to be responsible means to keep watch over those people. But is this linkage inherent, or is it circumstantial? If the execution layer were to change (say, hypothetically, some work being done by systems that don't fatigue or vary as humans do), would the nature of responsibility change with it? This is a question we will leave hanging for now. The key point here is that in the model as it exists today, oversight is virtually synonymous with accountability. If you carry real responsibility, you oversee; if you don't oversee, you're not really carrying responsibility in the traditional sense. This was a rational adaptation: it produced better outcomes and prevented countless mishaps. It also defined managerial work as we know it. To manage has meant to monitor and control, as much as to inspire or direct. Whether that must always be so is something we will explore later. For now, we recognize that oversight has been a cornerstone of making human-driven enterprises successful, the ever-watchful human element ensuring that other humans' work stays on course.

THE STABILITY THIS MODEL CREATED

The accumulated effect of all these design choices, hierarchy and roles, handoffs and oversight, has been to give organizations a remarkable degree of stability and scalability. By the late 20th century, it was routine for companies to grow to tens or even hundreds of thousands of employees, operating across multiple continents, all orchestrated through the structures we've discussed. That feat is truly historic. Never before could human efforts be coordinated on such a vast scale with such consistency. We should give credit where it is due: the classic organizational model works. It has allowed enterprises to meet ambitious goals reliably, to mass-produce products with minimal defects, to process millions of customer transactions correctly, and to comply with thick volumes of regulations and laws without collapsing. The model has proved durable across decades of economic cycles and technological changes. When people talk about "management 101" or "organizational best practices," they're often referring to principles that assume and reinforce this very structure. And for a long time, these principles delivered exactly what was promised: control, predictability, and repeatable success.

Because the model worked, organizations that mastered it were able to dominate their industries and expand. The structure provided real advantages in stability and scale. A clear chain of command meant decisions could be made in an orderly way and implemented throughout a large workforce. Formal procedures meant that whether a factory was in Chicago or Chennai, it operated with the same standards. Roles and training meant you could hire new people, slot them into the structure, and get consistent performance out of them after a period of onboarding. In short, the model was scalable. It was not fragilely dependent on any one superstar's genius; it was a machine that could be replicated, an "organization man" era ideal. The companies that epitomized the 20th century, from industrial giants like General Electric or Toyota to service behemoths like IBM or the U.S. federal agencies, all used bureaucracy and hierarchy to achieve feats of coordination that would awe a smaller, informally run firm. Weber observed that bureaucracy, for all its rigidity, is technically superior in precision, stability, and discipline, it creates an "efficient and predictable work environment" essential for handling large amounts of information and complex tasks. That predictability is exactly what allowed organizations to meet the expectations of shareholders, customers, and regulators year in and year out. You knew what you were getting, because the system ensured it.

Many of us built our careers and successes within this system. If you rose through the ranks, it was often because you demonstrated the ability to manage within the established structure, hitting targets, following procedures, exercising good judgment within your span of control. The organization rewarded those who could navigate its hierarchy and processes effectively. This is an important point: competence in the existing model is real competence. A plant manager who consistently delivers quality products on time by carefully adhering to and tweaking the factory's processes is demonstrating a mastery that deserves respect. A compliance officer who ensures a global firm has no major regulatory breaches for years is achieving something difficult, using the tools of oversight, training, and process diligently. These people succeeded because they learned how to make the machine run well. The machine, in turn, amplified their efforts. In a very tangible sense, the structure enabled scale but also created a stable career path for professionals, you knew that if you learned the ropes (how to get a project through the approval committee, how to manage a team within the company policies, etc.), you could be effective anywhere in the organization. The stability of the model and the way it rewarded mastery of that model are part of why it persisted. It wasn't just top-down imposition; generations of managers and employees internalized the logic and saw it pay off in their daily work.

We should also acknowledge how this model helped meet external expectations consistently. Governments and regulators, for instance, often look for stability and control in industries that impact the public. A hierarchical, process-rich organization can more credibly promise that it won't, say, ship contaminated food or misreport its finances, because it has checks and balances and accountability at every level. This was crucial for building public trust in the age of large corporations. Similarly, markets favor predictability, investors put a premium on companies that can reliably hit their numbers and avoid nasty surprises. The traditional organizational model excelled at that kind of predictability. Bureaucracy, in the positive sense, produced reliability at scale, which translated to economic and social success.

However, even as we celebrate the stability and scale the model created, we must remain aware that stability can sometimes mask accumulating strain. The very success of the system can make it hard to see its cracks. When changes in the business environment happen gradually, new technologies, shifts in workforce expectations, evolving competitive dynamics, a well-running organization might absorb the stress quietly, until one day the strain becomes evident. In a highly stable model, feedback loops that signal the need for change can be muted. After all, if quarterly results are still fine and operations hum along, why question the structure? Stability, in other words, can breed a kind of inertia. It's a bit like a long-standing building that shows hairline cracks only upon close inspection, while outwardly everything looks solid. Our organizational edifices have been tremendously solid, which is an achievement, but that solidity might make it harder to detect when foundational assumptions are shifting.

As we conclude this chapter, let's be clear: nothing we have discussed here suggests the model was wrong or foolish. On the contrary, it has been one of humanity's great managerial accomplishments to create organizations that harness our abilities and blunt our weaknesses so effectively. The point of laying this foundation is to truly understand why the current model exists and how much it has achieved. Only with that respect and insight can we thoughtfully examine whether anything is changing that might require a new approach. The stability and success of the traditional structure form the bedrock on which any new ideas must build. We stand on the shoulders of that giant.

WHAT THIS CHAPTER ESTABLISHES

We have now established the baseline for everything to come: modern organizations were deliberately built to compensate for human limits. The structures of roles, hierarchy, oversight, and process were not arbitrary, they were coherent, effective responses to predictable human constraints in judgment, capacity, and consistency. This design allowed companies to achieve an unprecedented level of scale and reliability. It shaped how responsibility functions (with oversight as a proxy for accountability), how work is divided (through specialization and handoffs), and how decisions are made (through hierarchical delegation). Nothing in this chapter argues that this model was misguided. In fact, the model is rational and proven in the context for which it was devised. These are the baseline assumptions: that humans are variable and finite, and thus we built organizations to be stable and vast by counterbalancing those traits.

By locking in this understanding, we can move forward in this book without having to re-explain why so many of our current practices exist. We can also avoid the trap of thinking "bureaucracy is bad" or "hierarchy is outdated" in a simplistic way. We now appreciate that those features had sound reasons behind them. As we explore potential shifts in later chapters, we will do so with a respect for the logic of the current system. After all, if something new is to replace or augment parts of this model, it will have to address the same fundamental issues of reliability, accountability, and scale that the old model managed so well.

Understanding this foundation is also crucial because it frames the key question: what if anything is changing about the underlying human constraints? The rest of the book will examine signs that the environment is shifting in ways that put strain on the old model. But before we jump into that, there is a natural next step. If organizations were built around human limits, then process, all those procedures and routines often derided as "red tape", was never just about slowing things down or exerting control. It was a protective mechanism, one of the most important we invented, to manage risk in a human-driven system. To fully appreciate how the current model might evolve, we first need to understand process on its own terms. Is all that process truly just inefficient bureaucracy, or is it serving a critical protective function? Before we consider changing it, we must understand why it exists.

Accordingly, in the next chapter we will turn our attention to process, not as an abstract concept, but as the living memory and risk management system of organizations. By seeing process as protection rather than pure overhead, we set the stage to decide what should be preserved, what might be modified, and what the real costs of change would be. Before we can question the burden of process, we must recognize the value it has provided. That is the logical next step now that we understand the organization we built for human limits.

2

PROCESS AS PROTECTION, NOT INEFFICIENCY

Organizational processes are often seen as bureaucratic hurdles that slow things down. Yet at their core, these processes are protective mechanisms designed to manage risk. Every checklist, approval flow, or sign-off procedure exists to prevent something from going wrong. In fact, process is one of the primary ways an organization compensates for the limits and variability of human performance. As discussed in Chapter 1, modern organizations were built around the reality that humans are inconsistent: prone to errors, lapses in attention, and divergent judgments. By introducing structured processes, companies aimed to reduce that inconsistency. They were not inventing bureaucracy for its own sake; they were creating a system to ensure reliability and safety in outcomes.

One key role of process is to reduce variance in how work is done. Consistency is a hallmark of risk management: if every employee follows the same proven steps, results become predictable and errors are less likely. In public writing and reporting, quality pioneer W. Edwards Deming stressed that uncontrolled variation is the enemy of good results. Processes implement that philosophy by standardizing tasks so that critical steps are not left to individual whim. A well-designed process catches predictable failure modes, the known points where things typically go wrong, and builds in checks or approvals at those points. For example, a process may require a second pair of eyes on a financial spreadsheet because experience shows complex calculations invite error if unchecked. Or it may mandate a legal review for a new contract because, without it, an enthusiastic manager might accidentally agree to risky terms. In all these cases, the process explicitly guards against known risks.

Process also makes accountability explicit. When work is structured into defined steps with designated owners, it's clear who is responsible for each outcome. This clarity is itself a safeguard. If a shipment requires both a preparer's and a supervisor's sign-off, then accountability is shared and transparent. No single individual can quietly cut corners without someone else's knowledge. And if an incident does occur, the organization can trace which step failed and correct it for the future. In this way, processes create an organizational memory of accountability where everyone knows the "rules of the road" and what they are responsible for, reducing the chance that a simple oversight snowballs into a serious issue.

Importantly, this protective role of process spans all functions and departments. Consider a few typical scenarios. In finance, companies institute controls such as requiring two approvals for large expenditures or payments. This isn't mere red tape; it ensures that no single person can inadvertently (or intentionally) make a costly error or fraudulent transaction. One person drafts the payment and another independently reviews and approves it, a classic risk mitigation known as separation of duties. Similarly, in the legal department, a contract review process might require lawyers to vet any agreement above a certain value or risk level. The goal is to catch unfavorable clauses or compliance issues that a non-specialist might miss, preventing predictable legal pitfalls. Human resources, too, relies on process: standardized hiring protocols help ensure new recruits are properly vetted so the company isn't exposed to discrimination claims or unqualified hires. An HR approval step for promotions or terminations provides a backstop against impulsive or biased personnel decisions. Across these examples, processes encode sensible precautions drawn from hard-earned experience about what can go wrong when such checks are absent.

On the operational front lines, checklists and standard procedures often make the difference between smooth performance and disaster. A classic example comes from aviation: early airplanes were simple enough that a pilot's memory sufficed, but as aircraft grew more complex, relying on memory became a liability. Public reporting on a 1935 crash of a new Boeing prototype noted that due to a forgotten step, pilots adopted the first pre-flight checklists. The effect was dramatic - using a simple checklist, crews subsequently flew the Boeing Model 299 many miles without incident. The checklist (a basic process document) turned a dangerously complex task into a manageable and safe operation. Similarly, in medicine, a routine surgical safety checklist has been shown to cut complications and deaths by roughly one third. This astonishing improvement wasn't achieved with new technology or more skill, but by instituting a standard process that catches human errors: verifying the correct patient, procedure, and instruments are in place, that antibiotics are given on time, and so forth. These real-world examples show that what might seem like mundane procedure (ticking off boxes on a list) can be a powerful barrier against catastrophe. In every case, process is the collective ritual of not letting our human guard down resulting in institutionalized caution.

These mechanisms were instituted because experience showed that doing things informally often invited unacceptable risk. In that sense, process embodies hard-won wisdom and enables high performance despite human fallibility. Thus, process merits respect as a coherent, effective response to real constraints.

Of course, adding all these safeguards and routines does not come without a price. Processes can indeed make work slower or less spontaneous, raising the question of what trade-offs have been made. To fully appreciate those trade-offs, we first need to understand the context in which process became so paramount. For most of modern organizational history, predictability was valued more highly than speed, and process was the means to achieve it. Understanding that historical bias toward predictability will shed light on why so many processes were designed the way they were and it will prepare us to consider how those priorities might be changing.

WHY PREDICTABILITY WAS MORE VALUABLE THAN SPEED

For most of the industrial age and well into the late 20th century, organizations operated in environments where predictability mattered more than speed. In these settings, an error could be far more costly than a slow pace. It was often wiser to be careful and correct than to be first and flawed. Consider the context in which many corporate processes evolved: industries like manufacturing, construction, healthcare, finance, and aviation. In such fields, a single mistake could have outsized consequences. A flaw in a batch of products might lead to an expensive recall or a safety hazard. A rushed decision on a construction site could result in structural failure. A misjudgment in healthcare could literally cost lives. In these conditions, being cautious and thorough was a rational response. Speed was valuable, yes, but avoiding catastrophe or irreparable error was invaluable.

Errors in traditional environments were often expensive, visible, or irreversible. If a bank released inaccurate financial statements, it could face regulatory penalties and a loss of public trust that might take years to rebuild. If an aerospace company skipped a testing step to save time, the failure of a component in flight would not only be public but possibly irrevocable. Likewise, a food company that rushed a new product to market without rigorous quality checks might have to pull products off shelves in a costly, reputation-damaging recall. In such examples, a slow and deliberate process functioned as a form of insurance. It ensured that before anything went out the door, be it a product, a financial report, or a medical procedure, multiple eyes had verified it, minimizing the chance of serious mistakes. The prevailing mentality was captured by adages like "measure twice, cut once" and "better safe than sorry." Going a bit slower, with more checks, was a conscious trade-off to guarantee things went right the first time.

This emphasis on predictability ties back to the limits of human execution we discussed earlier. Human beings working by hand or managing analog processes can only move so fast without increasing the risk of errors. There is a natural speed-accuracy trade-off in much of human work: try to do something twice as fast and you're likely to make more mistakes. Organizations understood this implicitly. They built process-heavy approaches to throttle pace at critical junctures, allowing time for review and correction. If a critical report was due, it was better to have a second reviewer take an extra day verifying the numbers than to send out a fast but flawed analysis. If a new pharmaceutical drug was under development, it was better to adhere to a painstaking clinical trial process than to rush it because regulators and the public would not forgive a harmful oversight, no matter how much time was "saved" by cutting corners. In short, the tolerance for risk was low, and process was the tool to systematically reduce risk. Speed was often treated as a secondary goal, something to pursue only once a high level of certainty and quality was assured.

Moreover, in many eras and industries, the external tempo of change was relatively slow by today's standards. Product cycles lasted years, major technological shifts were infrequent, and competitive landscapes were stable. In such a world, being the fastest wasn't as critical as being the most reliable. Customers and partners valued consistency; they needed to trust that a product or service would work as expected every time. A company that built a reputation for rock-solid reliability could dominate a market even by moving methodically. For example, a supplier of industrial equipment might take longer to fulfill an order because it ran extensive quality tests, but clients appreciated that diligence since it meant fewer breakdowns later. In banking, a cautious lending process meant fewer defaults and scandals because stakeholders preferred a bank that was prudent to one that was recklessly quick. Predictability was a hallmark of professionalism. An operation that ran like clockwork, even if that clock ticked a bit slowly, inspired confidence.

None of this is to say that speed was ignored. Rather, it was a conscious trade-off: some speed was willingly sacrificed to gain a lot of certainty. Leaders understood that trying to have both maximum speed and maximum certainty wasn't possible given human constraints and the tools of the time. If forced to choose, they chose to slow down to avoid costly mistakes. In an era before instant global communication and rapid digital iteration, a misstep could reverberate for a long time. Fixing errors was often more painful (or even impossible) than preventing them upfront. So the optimal strategy was to front-load work with process and scrutiny, ensuring reliability of outcomes.

Over time this bias toward predictability and risk avoidance set the tone for process design. Organizations essentially optimized for a world where the biggest threat was the unforeseen flaw, not the competitor who might beat you by a month. They optimized for complying with regulations, for quality control, for not making headlines with a scandal or a failure. Process was the vehicle for that optimization. If a process made outcomes more predictable e.g. less variance in quality, less chance of a lawsuit or accident, then it was deemed worth having, even if it added delay.

It's important to note that the balance between predictability and speed is not fixed for all time; it's a product of circumstances. In stable or high-risk environments, the balance tilted heavily toward predictability. If the environment changes, becoming faster-paced, more digital, more tolerant of minor errors in exchange for rapid innovation, then that balance might shift. For now, we acknowledge that for most of the period when our current organizational processes were born, predictability was king. Process became prevalent not because people didn't care about speed, but because they deliberately traded speed for assurance under the conditions of the time. They did so for sound reasons in context.

In short, process-heavy methods delivered what organizations valued most in that era: consistent, risk-free operations. They gave up some agility to get it. That trade-off was logical when a misstep could be ruinous and moving cautiously had little downside. Keeping this context in mind, we can better appreciate why process took the forms it did. Now we turn to another crucial aspect of process: how it serves as the repository of an organization's collective memory, preserving lessons learned over time.

PROCESS ENCODES INSTITUTIONAL MEMORY

Every process inside an organization tells a story. If you trace the origin of a given rule or procedure, you'll often find it was put in place as a direct response to a past lesson. In this way, process functions as a repository of institutional memory. It's the organization's way of remembering what has gone wrong before and ensuring those mistakes or near-misses are not repeated. Over years and decades, companies accumulate countless experiences of "what not to do." Process is how those memories are embedded into daily operations, long after the individuals involved have moved on. Procedures, checklists, and approval matrices serve as a living archive of hard-earned wisdom.

For example, imagine a company discovers that a certain type of contract once led to significant losses because a mid-level manager agreed to unfavorable terms. In the aftermath, leadership doesn't just scold those involved; they adjust the process. They might institute an approval threshold: henceforth, any contract above a certain dollar value or risk profile must be reviewed by a senior executive or the legal department. That new approval step is essentially a memorial of the painful lesson, formalized so the organization doesn't rely on individual recollection or vigilance alone. Now, even if everyone who remembers the bad contract eventually leaves the firm, the process will still be there to prevent a recurrence.

Such stories abound. Separation of duties in financial processes, requiring one person to authorize a payment and another to execute it, became a staple control because at some point someone with unchecked power committed fraud. The organization "remembered" that incident by splitting the roles, encoding that memory in policy. Similarly, many firms have escalation paths for handling problems: if a customer complaint or an internal issue isn't resolved within a certain time or at a certain level, it automatically escalates to higher management. Why? Usually because in the past, a small issue was allowed to fester at a low level until it became a major crisis. The company learned that lesson - don't let issues linger unseen - and built an escalation process to ensure visibility. Even something as mundane as a checklist for machine maintenance often has an origin story: perhaps once a critical machine failed because routine lubrication was neglected, so a new checklist item "inspect oil levels weekly" gets added. The next technician might not know the history, but by following the checklist they are implicitly guided by it.

Documentation requirements provide another illustration. Many industries mandate that decisions, tests, or communications be documented and archived. This can seem onerous: why write a report after every project or log every decision in a system? Often it's because at some point, lacking documentation caused a serious problem. Perhaps an organization faced a lawsuit and couldn't defend itself because it had no record of how a safety decision was made. Or an audit found discrepancies, and nobody could trace the source due to poor documentation. In response, the organization enshrined a new process: always document key actions and keep records for a set period. That process is effectively the memory of being burned by lack of information. It doesn't rely on each new employee's personal caution; it formalizes caution as standard practice.

When you view processes as stored experience, they stop looking like arbitrary red tape and start looking like collectively learned wisdom. This can be eye-opening for those on the front lines. An engineer might grumble about why there are so many code review steps before deployment, or a salesperson might question why legal approval takes so long for a custom deal. The answer is likely: because somewhere in the past, skipping those steps led to a bad outcome. That extra code review step might exist because a previous software release, lacking adequate review, caused a major outage. The protracted contract review might exist because a hastily drafted clause in an earlier deal triggered a lawsuit or fine. In other words, the process is there because something went wrong before. What feels like a hurdle today is actually the organization protecting itself based on yesterday's lessons.

Over time, these layers of learned experience accumulate. A mature organization's processes can be seen as a stratified sediment of incidents and responses: a near-miss here led to a new check there; a new regulation there led to a compliance procedure here; an audit finding resulted in an extra reporting requirement, and so on. This perspective fosters respect for process. Each rule or step was crafted by people who had a very good reason to put it in. They might have done so reluctantly, knowing it added burden, but recognizing that without it the organization was exposed to a risk it could no longer accept.

In this sense, process is the institutional memory that doesn't depend on any one person's memory. It allows an organization to scale and endure beyond the tenure of its founders or any individual team. New employees inherit a system that "remembers" the past for them. They don't have to personally experience every failure mode. If the system is well-designed, it guides them away from known pitfalls automatically. That is a huge advantage. It means an organization can maintain continuity and reliability even as people come and go.

However, institutional memory via process has a double edge. While it is a great strength that process preserves learning, it also means that process can outlive the conditions that created it. A rule implemented after one incident decades ago may still be followed faithfully long after the incident is irrelevant or the risk landscape has evolved. Organizations rarely forget but they rarely ask if everything in that memory is still needed. We won't argue here that old processes should be discarded because often it's better to over-remember than to forget a critical lesson, but it's worth noting that process, once established, tends to stick. It becomes part of the organization's DNA. In later chapters, we'll consider how changing environments can render some processes obsolete or overly costly. For now, the key takeaway is that process represents accumulated organizational learning. It should be understood and respected as such. Every checklist item, every approval chain, and every form in triplicate echoes some lesson learned. Seeing process this way helps us understand it not as capricious bureaucracy, but as the structural memory that has allowed organizations to avoid repeating mistakes and maintain continuity over time.

WHY PROCESS EXPANDS AS ORGANIZATIONS SCALE

As an organization grows in size and complexity, informal ways of working start to break down. A small team of five or ten people can coordinate through constant communication and mutual awareness. They don't need much formal process because everyone is in the loop by default. They are often in the same room, sharing information in real time. Trust is high, roles are fluid, and adjustments happen on the fly. However, when a company becomes fifty people, then five hundred, then five thousand, that organic coordination cannot scale. Process fills the gaps left by the loss of shared context and direct trust.

Imagine a startup that has grown into a mid-sized enterprise. In the early days, if a customer had a problem, the CEO might hear about it immediately and the team would huddle to fix it by day's end. There were no formal tickets, no escalation matrices. Fast forward: the company now has dedicated support staff, separate product teams, a sales department, and regional offices. The CEO is no longer on every email thread. Suddenly, when a customer issue arises, it's not obvious who should handle it or who needs to know. Without some structure, one group might assume another group is taking care of it and thus no one addresses it. To prevent such breakdowns, the company introduces processes for coordination: a support ticket system with defined response times, an escalation path if an issue isn't resolved, regular cross-department check-ins to ensure information flows. These processes create an artificial but necessary way to mimic the tight communication that used to happen naturally in the small team.

The larger an organization gets, the more it must rely on explicit rules and routines instead of implicit understanding. In a tight-knit crew, everyone might know "Alice handles client contracts" simply through daily interaction. But in a huge company, many people dealing with client contracts may never have met Alice or even know she exists. So the company establishes a formal process: all large client contracts must route through the Contracts Department (where Alice works) for approval. That's scale at work. It's not that someone arbitrarily decided to "add bureaucracy." Scale forces a shift from personalized coordination to standardized process. When hundreds or thousands of employees must collaborate, processes ensure they do so coherently. Standard forms, request channels, approval hierarchies, and documented workflows become the glue holding a complex operation together.

People often attribute heavy process in big organizations to leadership style ("Our management is too controlling" or "We've become too cautious"). In truth, while leadership influences the flavor of processes, the existence of extensive process is largely a structural consequence of growth. If you took the most freewheeling startup leaders and put them in charge of a 50,000-person enterprise, they would eventually implement many of the same processes as any large enterprise. They'd have to because without those processes, the organization would descend into chaos or inconsistency. The challenges of scale (keeping everyone aligned, ensuring quality across outputs, complying with laws in many jurisdictions, onboarding waves of new employees) push even innovative companies toward formal structures. That's why big organizations, whether corporations, governments, or armies, all end up relying on hierarchy and process. It's not a failure of imagination; it's a reflection of common constraints.

Thus, process growth is not a moral failing of an organization; it's almost a law of nature. With growth, additional process is as predictable as increased gravity on a larger planet. Each new division needs some rules of engagement with others. Each new initiative spawns guidelines to integrate with everything else. Over time, the accumulation of these rules can seem bewildering from the ground, but from above it had a logic: it was solving coordination problems at a larger scale.

However, while the rationale for process remains valid as an organization scales, the cost profile of process changes. A two-step approval in a small team might add negligible delay; that same approval, when thousands of such decisions are occurring across multiple layers, can introduce significant lead time and overhead. So, even though each process addition at scale made sense in isolation, collectively they can impose a drag on agility. In other words, scale increases the density of processes and their interdependencies, which can amplify their costs. We'll examine these hidden costs next. For now, the key point is that as organizations grew, processes expanded naturally as a supportive scaffold. That expansion made large-scale coordination and reliability possible - an undeniable benefit - even as it set the stage for new challenges related to complexity and speed.

THE HIDDEN COSTS OF PROTECTIVE PROCESS

We have emphasized the virtues of process and how it prevents errors, preserves knowledge, and enables scale. Equally, we must acknowledge that these protective processes come with costs. Every checkpoint, approval, and control introduces some friction. Individually the impact might be minor, but collectively the burden can grow. Process reduces certain risks but also introduces new challenges and even new kinds of risk.

One cost is increased coordination effort. Processes often involve more people in a task (for good reason, like double-checking work or splitting authority). However, getting those people aligned takes time and energy. A project that now requires five approvals (where one manager once sufficed) means scheduling meetings, circulating documents for review, fielding questions from approvers, and chasing signatures. Coordination can become a project unto itself. In large organizations, employees may spend a substantial portion of their week on internal processes: updating reports, obtaining sign-offs, attending governance meetings. That's time not spent on direct value creation; it's the price paid for safety and compliance. There's a trade-off between doing the work and managing the work. As processes proliferate, people can find themselves devoting as much effort to navigating the process as to executing their primary job.

Another cost is slower decision-making and reduced agility. By design, protective processes put checks in place that delay action until conditions are met. This can be lifesaving in a hazardous context (no one wants a nuclear plant operator skipping a safety checklist). But in a fast-moving business, those delays can mean missed opportunities. If approving a marketing campaign takes six weeks due to compliance reviews, legal reviews, committee meetings, and executive sign-off, a market trend might pass by the time the campaign finally launches. Front-line employees might see what needs doing immediately, but process enforces patience and thoroughness at the expense of speed. The result is that organizations become less responsive. Competitors or events may outpace them while they are still navigating the internal maze for approval. Historically, that trade-off (slower but safer) made sense; it can become a liability when speed itself is safety (for example, responding quickly to a cybersecurity threat or a sudden market shift).

A further cost is reduced local autonomy and empowerment. The more process dictates what must happen, the less room individuals have to use their judgment. If a customer service employee must follow a strict script and get managerial approval for any exception, they may feel powerless to solve a customer's unique problem on the spot. They become a cog in the machine, executing steps without the authority to adapt. This can dampen motivation, as people feel their creativity and expertise are underused. It can also transfer risk in an unexpected way: if everyone is following the process to the letter, people might not speak up or intervene when something outside the process is going wrong. A subtle mindset can take hold: "I did everything the procedure said, so if something's off, it's not my fault." The intention of process is to clarify responsibility, but the effect can be that individuals stop exercising initiative, assuming the process will catch any issues.

There's also the issue of handling exceptions. No matter how detailed a process is, reality will produce cases that don't fit neatly. Procedural organizations often create secondary processes for these: special approvals or committees for non-standard requests. This means alongside the normal workflow, there's a parallel workflow for exceptions. Managing those can be cumbersome. Sometimes work just stalls because the process has no path for an unusual scenario. Or people quietly bypass rules to get things done, introducing risk because now actions happen off the record. In a heavy-process culture, exceptions can become normalized in a paradoxical way: everyone knows that "officially" you do X, but in practice people often do Y as a workaround when X is too rigid. That workaround culture is itself risky. It means the formal process might be masking a lot of informal activity not subject to intended controls.

All these costs - slower decisions, more effort on coordination, less empowerment, difficulty with exceptions - are inherent trade-offs in a process-rich organization. They are not arguments to abandon process entirely, but they must be acknowledged. In Chapter 1, we noted that compensating for human limitations has rising costs as complexity grows. We see that plainly here: the safer and more controlled you make a system, the more time and resources that control tends to consume. There is no free lunch. Every extra review step that might catch an error is also a step that could slow a response. Every additional form that ensures completeness also takes time to fill out.

Crucially, pointing out these costs is not a moral indictment of process. Earlier, we established why processes exist and the crucial protection they provide. Those benefits remain real. But balance is key. An organization must continuously weigh the diminishing returns of adding more process against the risks being mitigated. A lightweight process might miss some issues; a heavyweight one will catch more but at increasing expense. There comes a point where protective mechanisms can start to create new risks, for example, a company might become so slow it loses market opportunities, or talented employees might leave because they feel stifled. In extreme cases, a process intended to prevent failure can contribute to failure by making the organization too rigid to adapt to a novel challenge.

The challenge, then, is finding the right equilibrium. Many organizations struggle with how to be safe but not stagnant and how to maintain control without suffocating creativity and speed. The takeaway here is that every process carries trade-offs. Its protective value must always be weighed against its hidden costs. If those costs quietly grow to outweigh the benefits, then the process is no longer truly protective but instead it's undermining the very goals it was meant to secure. Next we'll discuss why, when things go wrong or feel sluggish, process often gets the blame. But as we'll see, blaming process alone can be misleading if we don't consider the deeper causes at play.

WHY PROCESS IS OFTEN MISDIAGNOSED

When organizations struggle - slow product launches, high costs, frustrated employees - process is often the first suspect. It's easy to see why. Process is highly visible: the layers of approvals, the forms to fill out, the meetings to get anything decided. People point to tangible frustrations like "we have too many sign-offs" or "this request sat in three committees for months." Bureaucracy becomes a convenient villain when outcomes lag or teams feel bogged down. However, this tendency to blame process can be a misdiagnosis. Often, the angst over "too much process" is actually a symptom of deeper structural issues or of processes that haven't been updated to match a changing environment.

One reason process gets blamed is simply because it's there and an obvious target for dissatisfaction. Unlike a fuzzy issue like strategic misalignment, bureaucracy feels concrete and fixable: just cut it! And indeed, trimming genuinely unnecessary process can bring relief. But process is not always the true root cause of an organization's woes. It's the surface irritation everyone feels, but the underlying causes may lie elsewhere. Perhaps the business has grown more complex but hasn't upgraded its systems, so old processes now feel unbearably slow. Maybe decision-making authority is unclear, so everything defaults to multi-layer approval. The processes may be cumbersome because the organization's structure or strategy is flawed.

For example, a new CEO might arrive and decry the 1,300 pages of internal rules and a dozen layers of hierarchy choking a company. Certainly, that sounds excessive. But how did it get that way? In a firm like that, decades of global operations, regulatory pressures, and risk management needs accreted those layers. Blaming "bureaucracy" alone doesn't provide a strategy for improvement. In fact, as one analyst noted, much of that structure exists to coordinate extremely complex work at scale. Middle managers and internal rules often play critical roles in keeping the machine running. If you simply slash them without a plan, you might remove waste or you might remove the framework holding a complex enterprise together.

Often, complaints about process are stand-ins for other frustrations. Employees might actually be upset that priorities keep shifting, or that the company's risk tolerance is so low that every minor decision requires approval. These are deeper issues. Piling on more process can be a bandage if leadership hasn't addressed the real problem. Conversely, cutting process without addressing the real problem can backfire. If, for example, a company's strategic direction is unclear, it will create lots of approval loops simply because no one is sure what should be prioritized so every decision gets escalated. The root cause there is a lack of strategic clarity. Eliminating approval steps won't magically create alignment; it might just lead to uncoordinated decisions. Similarly, if a company's culture is very risk-averse, it might overlay process as a safety net. Remove those layers without changing the culture, and you may simply expose the company to risks it isn't prepared to manage, or paralyze employees who have been conditioned to rely on those checks.

In essence, process problems are often proxy problems. They signal something else - complexity, misalignment, outdated assumptions, or resource gaps - that needs attention. Before rushing to "cut bureaucracy," it's crucial to ask what each burdensome process is actually doing and why it exists. For example, if people complain of too many meetings and forms, is it because decision rights are unclear or data is siloed? If so, the deeper fix might be clarifying authority or improving information flow, not simply canceling meetings or forms.

The key point is that blaming process alone for an organization's malaise is too simplistic. Yes, cumbersome processes can drain energy and slow things down, as we've seen. But they usually got there for rational reasons. Before we tear them down, we have to ensure we understand those reasons. Otherwise, we risk reintroducing old failures or stripping away safeguards we actually need. The better approach is to remedy the underlying conditions rather than just attacking the visible symptoms of bureaucracy.

Thus, before reducing or redesigning any major process, it's necessary to understand what conditions made it necessary in the first place. What failure was it preventing? What human limitation was it addressing? If those factors still exist, simply cutting the process may do more harm than good. But if those factors have changed or can be addressed in a better way, then the process is a good candidate for rethinking. Approaching things with this mindset ensures that any simplification or innovation isn't just a reckless teardown, but a thoughtful evolution. It's about solving the real problems rather than just removing visible irritants.

WHAT THIS CHAPTER ESTABLISHES

This chapter reframed process not as pointless bureaucracy, but as a protective system that organizations put in place for good reasons. We explored how modern processes arose to manage human limits, to capture institutional learning from past events, and to coordinate work at scale. In all these respects, process has real value: it makes outcomes more predictable, prevents known mistakes, and allows large groups of people to function together reliably. Many readers likely recognized their own workplaces in examples of financial controls, legal sign-offs, HR procedures, and operational checklists and hopefully saw the rational intent behind them.

At the same time, we took a clear-eyed look at the costs of process. We cataloged how process can slow things down, create extra work, reduce flexibility, and sometimes encourage a "check the box" mentality. These downsides are real trade-offs that every organization must manage. A key insight is that both the value and the cost of process are genuine. Process brings safety, consistency, and memory but it also brings overhead, delay, and rigidity. Understanding this duality is essential for anyone evaluating or changing how an organization operates.

The important thing this chapter establishes is context and respect. Nothing here suggests that the solution is to simply "abolish process." On the contrary, we've shown that processes exist for reasons that deserve respect. They were coherent responses to real constraints and risks. But we've also set the stage to ask whether those constraints and risks are evolving. The world doesn't stand still, and an organization's protective measures may need reexamination as conditions change.

We now have a shared mental model: process is protection and it has a purpose and a price. Equipped with that understanding, we can talk about how the world around these processes might be shifting, and what that means for the way organizations operate. Before jumping into the future, however, the next chapter will focus on the human side of the equation: why mastering the current process-driven model has been a genuine form of competence, and why that mastery still matters. After all, if processes and structures were designed to manage risk and achieve stability, then mastery of this model - knowing how to navigate and lead within these processes - has been a key to organizational success. Appreciating that will ensure that any evolution of the model builds on strengths rather than discards them.

3

WHY MASTERY OF THIS MODEL STILL MATTERS

Operating effectively inside today's organizational model requires genuine skill. This is not a trivial observation. Large companies and institutions run on structured processes, hierarchies, and careful controls, and navigating that landscape successfully is real work. It takes process fluency to understand how things get done: knowing the sequence of steps, the forms to fill, the checkpoints to clear. It takes approval navigation, the knack for getting proposals through layers of sign-offs without stalling. It takes risk triage, being able to distinguish the issues that need immediate escalation from those that can be safely handled at lower levels. It takes deft exception handling, because no process covers every situation and someone has to decide when to bend a rule or seek a one-time exemption. And it takes influence through structure, the ability to work within formal decision channels and still shape outcomes. None of these competencies are innate or automatic; they are learned through experience and careful study of "how things work around here."

Indeed, operating within a complex organization is often as challenging as the substantive work the organization does. A new employee quickly discovers that success isn't just about doing their functional job (be it marketing, engineering, finance, etc.), but about understanding the system that surrounds that job. The hierarchy, the process, the approvals, and the accountabilities form an ecosystem that must be skillfully managed. If you've spent years learning how to get a budget request through all the reviews, or how to navigate the chain of command to resolve a customer issue, you have developed a form of expertise. In a very real sense, you've become fluent in the "language" of your organization's model, a language that outsiders often don't even know exists.

Crucially, this competence is not an archaic or obsolete talent. It's sometimes fashionable to mock bureaucracy or to imply that people who thrive in it are merely "playing politics." But the truth is that mastery of the current system is substantive. Consider the daily reality of managing within a regulated industry or a public agency: there are legal compliance steps that, if skipped, could shut down the business; there are quality control processes that save customers from defects or hazards; there are approval matrices that ensure senior leaders are aware of big risks. Guiding a project to completion under these conditions, without triggering failures or violations, demands planning, foresight, and steady leadership. Far from being accidental or purely political, success in such an environment reflects deep understanding. As one expert noted in the context of government, bureaucracy isn't just red tape to be mocked, it's a system of rules and routines developed over time to manage risk, reduce bias, and deliver consistent results. In other words, the structures so many of us work within were designed to produce stability and reliability, and it takes real know-how to make them function as intended.

Mastery of this model also means understanding the flow of decisions and the management of risk in a human-limited environment. Chapter 1 explained how organizations were built for human limits, and Chapter 2 showed process as a form of protection. Within that context, to "know the ropes" is to understand how risk is managed at each juncture, who needs to sign off because they carry a certain accountability, which checkpoints are there to catch predictable failure modes, how a minor exception can snowball if not contained. It means knowing how decisions move: for example, recognizing that an informal nod from a senior stakeholder can smooth a formal approval later, or that a detailed backup memo may be needed to satisfy an oversight committee even if it's not explicitly asked for. It means knowing how outcomes are stabilized: perhaps by double-checking that a backup system is in place or that multiple departments have coordinated before a launch, because you've seen what happens when they haven't.

These skills aren't written on any diploma, yet careers have been built on them. And importantly, they deserve respect. The company may have an org chart and a published process manual, but the true day-to-day functioning relies on professionals who can interpret and apply those structures in messy real life. Nothing about that is easy or automatic. In fact, many bright newcomers struggle until they acquire this implicit knowledge. Fluency in the current model is a form of expertise. It's a competency developed through effort, often via mentorship from seasoned colleagues or through hard knocks when something falls through a gap in the process.

Because this mastery is real, it naturally breeds a sense of confidence in the status quo. When you know how to work the machine, you also understand its value and its logic. That in turn can create understandable skepticism toward those who claim the whole machine is broken or needs to be reinvented. After all, if you've spent years learning how to make the system deliver results, you know it can deliver results, and you might bristle at the suggestion that your success was due merely to "playing politics" or that it's irrelevant in a new scheme. This is a critical point: the legitimacy of current competence can become a source of resistance to change. People push back on radical change not just out of self-interest, but out of a genuine belief that one shouldn't carelessly discard a well-honed system. In their experience, the system exists for good reasons, and they've proven that they can excel within it. So any proposal to upend it must overcome an implicit question: "Are you sure we're not throwing away something valuable?" In short, the very reality of mastery means the current model has defenders who are both intellectually and emotionally invested. This is not obstructionism; it is the rational hesitance of experts who know the terrain. It comes from pride in hard-won competence. We will soon explore how this dynamic affects openness to change, but first, let's delve deeper into what that competence really entails on a human level.

EXPERIENCE IS EMBEDDED IN HOW WORK GETS DONE

Much of the expertise that keeps organizations running is not abstract theory, it lives in the tacit knowledge of experienced individuals and in the routines of teams. Over years and decades, people accumulate a nuanced sense of "how things are done around here." This knowledge is often procedural and situation-specific, rather than something you could just look up in a manual. Think of a veteran project manager who can sense which preliminary meetings will smooth the path for a proposal, or an operations supervisor who instinctively double-checks a particular gauge because they recall a near-miss years ago when it malfunctioned. These are insights that typically aren't written down formally; they reside in human memory and organizational habit.

One way to appreciate this is to look at how seasoned practitioners navigate process. An experienced employee often knows which steps truly matter and which are more ritual than substance. For example, they might know that a certain sign-off is essentially a formality, perhaps a senior executive's "approval" that is always given as long as a mid-level committee has vetted the details. Conversely, they know which reviews are essential, say, a legal compliance review that, if skipped or rushed, could expose the organization to serious risk. They learn where judgment must be applied carefully: maybe an expense policy has gray areas, and they've learned when to seek explicit exception versus when to use discretion. They even know which approvals are symbolic, maybe a particular department head likes to be asked out of courtesy, even though they rarely veto anything, and failing to involve them would create ill will. This kind of intricate understanding of process and structure isn't in the flowchart; it's learned by observing outcomes over time.

Such knowledge is often earned through repetition and exposure to failure. Every time something goes wrong, a project overruns, an audit flags an issue, a client complaint escalates, it leaves a trace in the organizational memory. Sometimes that trace is explicit, as in a new checklist item or an amended procedure. But often it's informal: people remember the scare and adjust their behavior to prevent a repeat. In Chapter 2, we discussed how process can encode institutional memory, serving as a repository of lessons learned. Here we see the human side of that: experienced people become walking repositories of institutional memory. A senior operations director might say, "We require that second safety inspection because five years ago we had a serious incident, I was here when it happened." Or a finance director might enforce a particular review because, "Trust me, we tried skipping it in the past and paid for it." There's a saying in the world of safety and compliance that "every rule is written in blood," meaning every regulation or procedure exists because someone got hurt or something went badly wrong, and a rule was invented to prevent it from happening again. In a corporate context, it may not always be literal blood, but the principle holds: behind many formal processes lies a backstory of hard-earned experience.

This depth of experience is inseparable from the current structure. It's not easily transferable outside the context in which it was learned. If a supply chain manager has spent 20 years mastering the intricacies of their company's global logistics process, that expertise isn't just "supply chain knowledge" in general, it's deeply tied to how this company operates: the quirks of its systems, the relationships with its particular suppliers, the internal culture around risk. If you pluck that manager out and drop them into a vastly different environment, some of their skill will carry over, of course, but much of their finely tuned intuition might not apply. They would have to learn a new landscape from scratch. This illustrates a key point: experience in an organization is context-rich. It's one reason companies value long tenure, because those individuals carry a map of the organizational minefields and hidden pathways that simply can't be taught overnight. The current model, with all its processes and structures, is the scaffolding on which that experience is built.

Another aspect of this embedded experience is how it gets shared (or doesn't) across the organization. Often, it's passed down person to person. New hires learn the real ropes not just from orientation manuals, but from mentors or team leads who show them the informal tricks of the trade. In this way, the mastery of process and structure is a kind of apprenticeship system. Over time, almost subconsciously, people internalize the values and priorities encoded in the processes. For instance, if every time you try to rush a step you get pushback that "we need to do the due diligence," you learn that thoroughness is prized over speed. If ignoring a small anomaly results in a stern lecture about following procedures, you learn that consistency is critical. Thus, the culture and the process together educate employees about "how we do things here." And those who have absorbed these lessons most thoroughly often become the culture carriers for the next generation of staff.

It's also worth noting that this tacit knowledge is often more procedural than theoretical. People may not articulate the grand rationale behind a rule, but they know when to apply it and how. They might say, "In situation X, make sure you loop in the risk team early, it'll save you trouble later," without necessarily describing the full risk management theory at work. In practice, that's fine: organizations need doers who understand the practice deeply, not just the theory. But it does mean that a lot of the organization's wisdom resides in habits and norms rather than written doctrine.

All of this underscores why the current model's mastery is valuable: it embodies the organization's cumulative learning. The processes and the people's know-how have co-evolved. Change the processes too abruptly, and you risk severing the link to that memory. Yet, as we'll see later, keeping processes forever can also lock in old assumptions. For now, we acknowledge that when work has been done essentially the same way for a long time, the people doing it have developed an intuitive expertise that can't be replicated easily. When the formal structure shifts, that experience doesn't disappear, but it may need to express itself differently. A veteran's instincts will still be there, but they might have to re-map those instincts onto a new set of procedures or tools. That can be challenging, it's like asking someone fluent in one dialect to suddenly operate in another. They can do it, but some nuance is inevitably lost in translation, at least initially.

By recognizing that experience is deeply embedded in the current model, we set the stage for an important consideration: any evolution in the model must find a way to carry forward this hard-won knowledge. Seasoned professionals have accumulated wisdom that a new system should leverage, not ignore. Part of respecting mastery is understanding the wealth contained in those tacit understandings. As we affirm the value of that embedded expertise, we also start to see why sudden changes can be so disruptive, they don't just rearrange workflows; they threaten to dislodge the accumulated wisdom of how to make those workflows successful. And that naturally makes people cautious about change. Before we get to how to navigate that, let's examine why organizations have historically rewarded those who attain this mastery in the first place.

WHY ORGANIZATIONS REWARD THIS MASTERY

If you look at who rises through the ranks in traditional organizations, you'll often find it's the people who have demonstrated they can work reliably within the existing system. This is not a coincidence; it's by design. Organizations, especially large ones, value predictability and risk avoidance. The entire structure is typically set up to prioritize consistent outcomes over dramatic swings. Therefore, individuals who can "move work through the process" smoothly, without causing crises, become linchpins of the enterprise, and the enterprise rewards them in turn.

Think about what a promotion or increased responsibility signifies in a conventional organization. It usually means the person has shown good judgment, delivered results that met the organization's standards, and managed their risks well. Perhaps they have overseen projects that stayed within budget and complied with all required regulations. Perhaps they have a track record of no major failures on their watch. These are exactly the outcomes the current model is engineered to achieve: stability, compliance, steady progress. So it makes sense that the incentive and career structures align with those outcomes. People who embody the system's values of caution and thoroughness tend to advance. In many companies, you might hear phrases like "a safe pair of hands" used to describe a desirable leader, someone who won't drop the ball or steer into chaos. Promotions, in other words, often reflect trust that the person will uphold the existing system's reliability.

There is a practical rationale here. Large organizations have a lot to lose if something goes wrong: revenue, reputation, even lives in some cases. The whole point of building layers of process (as we saw in Chapter 2) was to prevent mistakes and protect against downside risks. So when someone proves adept at managing within those protective layers, hitting their numbers while following all the rules, the organization naturally wants them to take on more responsibility. They have shown they won't recklessly gamble with the company's well-being. In a sense, these individuals become guardians of predictability. They perpetuate the success formula that has worked historically. Their reward (career progression, pay, authority) is society's signal that their skill set is valued.

This dynamic can be observed in virtually every function and industry. A sales manager who consistently delivers forecasts without nasty surprises, all while adhering to pricing and compliance rules, is often promoted to a regional director. An engineer who designs products that meet quality standards and pass all regulatory hurdles (even if it takes a bit longer) will be trusted with larger projects. A finance controller who never has accounting irregularities and passes every audit will likely climb to CFO over one who delivers flashier profits at the expense of bending rules. The pattern is clear: reliability and system fluency are rewarded.

It's important to note that this isn't simply about conservative personalities rising over bold ones. It's about what the organization needs and measures. For much of modern organizational history, as earlier chapters discussed, the environment favored those who could ensure stability. Many industries operated in conditions where errors or compliance failures were extremely costly, and speed was secondary. In such environments, the heroes were the people who could navigate bureaucracy expertly to avoid landmines. They knew how to get a new drug through FDA approval, or how to implement an IT system without breaking any data privacy laws, or how to manage a factory with zero safety incidents. These achievements are deeply admired internally, and for good reason. They keep the enterprise alive and well. It's no wonder the career paths and authority structures bent toward those who mastered these competencies.

Over time, of course, this creates a self-reinforcing culture. The leaders at the top usually come from this mold of mastery, so they set expectations for the next generation in a similar vein. New employees see what behaviors are rewarded and adjust accordingly. An organizational norm takes hold: do things the established, safe way if you want to get ahead. The positive side of this is consistency and clarity, people know what it takes to be successful. The potential downside, which we'll explore in later chapters, is that if the environment shifts to value different behaviors (say, more agility or innovation), the incentive system might lag behind that shift. For now, though, the key point is that the current model intentionally rewards those who uphold it. This is not a mistake or a failing of the organization; it's actually evidence that the system is coherent. The structures, processes, and incentives are all aligned towards the same goal: keep things working smoothly and avoid unpleasant surprises.

It's also worth acknowledging how personal and meaningful these rewards are to those who receive them. If you've spent 15 or 20 years rising through a company by mastering its way of working, your identity is likely wrapped up in being one of the "people who can get things done here." You might justifiably take pride in being that go-to person who understands how to navigate the maze. Your promotions and raises are tangible recognition of your competence. So when someone comes along and suggests a dramatic change, perhaps implying that a lot of this careful work is "old-fashioned" or inefficient, it can feel like not just a critique of the system, but a critique of you. In many cases, it is not paranoia to hear it that way; sometimes the change advocates do downplay the old expertise (unintentionally or not). That is why change can feel threatening: it appears to devalue hard-earned competence and, by extension, the people who embody that competence.

This emotional reality is crucial for any change effort to account for, but again, it starts from a rational place. The organization told these individuals, through years of rewards, "Your mastery matters; we need it." And that message was true. It remains true even as we contemplate changes. The mastery of the current model remains necessary in many respects, companies cannot simply throw away processes and expect overnight excellence in a new mode. So as we continue, keep in mind this dual perspective: the existing model's masters have been rightly celebrated for their contributions, and any evolution must contend with the fact that the old incentives and the new needs might diverge. That tension, between what the organization has historically rewarded and what it might need to reward going forward, will surface as a key theme. But rather than jump ahead, let's reinforce one more aspect of why the current mastery is so essential: it has enabled organizations to reach the scale and complexity we see today.

MASTERY ENABLES SCALE, NOT JUST CONTROL

When people critique bureaucratic organizations, they often focus on control, the rules, the oversight, the feeling of being constrained. But an overlooked truth is that the mastery of the bureaucratic model is what enabled organizations to scale to massive size in the first place. The structures and processes of the modern corporation weren't developed just to rein people in; they were developed to allow a company to grow from ten people to ten thousand and still function. Mastery of this model by its managers and employees is what makes that possible. In other words, the system isn't only about preventing chaos; it's also about making large-scale coordination feasible.

Consider a simple question: How can a product or service remain consistent across dozens of locations and thousands of employees? Part of the answer is having standard processes, yes, but the other part is having people who know how to implement and maintain those processes across diverse contexts. For example, a fast-food chain can give every franchise the same operations manual, but it relies on regional managers and store managers who master that system to actually deliver a uniform customer experience worldwide. Without those skilled managers, the manuals alone would not suffice. Or take a multinational tech company rolling out a new compliance protocol in response to a global data privacy law: the policy might be centrally designed, but it works only because local compliance officers in each country understand how to integrate it into their existing workflows, train their staff, and monitor adherence. These are people expertly operating within a structured model, ensuring that a change is absorbed without breaking the machine.

Scale amplifies everything, both the good and the bad. We often talk about economies of scale on the cost side, but there's also a kind of "economy of scale" of competence: a well-designed process, executed by competent people, can replicate a success many times over. That's a fundamental reason companies build standard processes, so that you don't have to reinvent the wheel for every new office or every new hire. However, the existence of a standard alone isn't enough; you need the human element to carry it out correctly. If one office interprets the procedure differently or an employee skips steps out of ignorance, the consistency is lost. Therefore, organizations rely on masters of the model to propagate and uphold the standards. They are the links that keep the chain strong across vast spans of operation.

As mentioned earlier in this book, Max Weber, the sociologist who studied bureaucracy, famously argued over a century ago that bureaucracy is the most efficient way to manage a large organization, that it allows "precision, speed, unambiguity... continuity, and unity" at an optimal level. While our world has evolved since Weber's time, the core insight remains: formal structure and clear processes allow big systems to run. And those benefits are not automatic; they are realized through the day-to-day mastery of countless individuals. In a strictly bureaucratic administration, as Weber noted, things like knowledge of files (information), reduction of friction, and clear subordination can reach a high point, but only if people work to make it so. The reason a 50,000-person company can deliver reliable products year after year is because legions of its employees know how to do their part within the grand scheme and how to connect their part to others'.

It's also essential to dispel the notion that mastery in this model is about mindless compliance. On the contrary, judgment and situational awareness are key components of effective bureaucracy management. A common misconception is that bureaucracies run on autopilot through rules alone. In reality, rules and processes cover the baseline, but human judgment fills in all the gaps and resolves the conflicts when rules collide or don't exactly fit. For example, a skilled manager knows how to prioritize when ten urgent requests all meet the official criteria but cannot all be done at once; they apply judgment to sequence work in a way that best serves the organization's true priorities. Or consider situational awareness: during a crisis or a special initiative, an expert operator might temporarily simplify or streamline parts of the process (with proper authorization) to meet a deadline, and then restore the normal process after, effectively flexing the structure without breaking it. This kind of flexibility is itself a mark of mastery: knowing when and how to adjust the rigid system carefully to handle special cases, and doing so in a controlled manner. It's reminiscent of an engineer knowing exactly how much stress a bridge can take and when to open or close lanes to keep it safe under unusual loads.

In fact, the best people within the current model are not those who mindlessly say "no" or robotically follow checklists. They are those who deeply understand the intent behind the rules and can thus implement them intelligently. They achieve consistency and efficiency by not overburdening the system with pointless bureaucracy, even as they uphold necessary bureaucracy. In large organizations, we often see internal innovators who find ways to make processes more efficient without violating any controls. They succeed because their credibility within the system is high, colleagues know they won't compromise core principles, which gives them latitude to suggest improvements. This is mastery at a higher level: not just following the process, but refining it from within, in a way that still respects its protective purpose.

All these abilities have enabled organizations to grow and span continents and decades. If you imagine trying to run a corporation of today's size with a totally ad-hoc, informal approach, it simply wouldn't work. You'd have confusion, duplicated effort, and things falling through cracks constantly. It was the evolution of structured roles, hierarchies, procedures, and trained managers that allowed enterprises to become so large and yet not collapse under their own weight. Mastery of the model by people is what converts a bureaucratic diagram into a living, breathing enterprise that can accomplish great feats, building airplanes, running global supply chains, administering public programs to millions, you name it.

However, with scale also comes complexity and cost. As organizations reach extreme sizes, the model that enabled that scale can start to strain under its own success. Coordination across thousands of people incurs overhead, meetings, reports, audits, layers of approval. Mastery is still making it possible to manage, but the effort to maintain mastery grows. We can see hints of this when seasoned managers talk about how much time they spend "managing the process" versus doing what they consider the "real work." At a small scale, process is a minor tax; at a huge scale, process can become a major line item in the cost structure of the company (in terms of time and money). This doesn't mean the model is broken, it's the natural price of having achieved such scale in the first place. But it does mean that the very triumph of the current model (large, stable operations) sets up the next challenge: how to handle coordination costs and complexity that keep rising.

The important takeaway for now is positive: we should appreciate that the model's mastery has not just been about enforcing rules for their own sake; it has been about enabling growth and reach that would otherwise be impossible. The leaders and experts who are deeply fluent in the system have been the ones extending the organization's capabilities safely into new arenas. The model is not just about control, it's about making large systems workable. This perspective will be vital as we consider changes. We aren't looking to abandon the benefits of scale or the wisdom embedded in the current system. Any future evolution must strive to keep what has made large-scale success possible. As we transition to examining where the model shows strain, remember that scale itself changes the game: the larger and more complex the organization, the more effort it takes to coordinate and maintain consistency. Mastery can handle growing coordination demands up to a point, but beyond that point, even masters start to feel the system's weight. We will explore those limits soon. Before that, though, let's address a critical caution: the dangers of trying to dismiss or bypass the existing model without due respect.

WHY DISMISSING THE MODEL CREATES RISK

Given that we've established how much competence is baked into the current system, it follows that simply ignoring or bypassing that system can be perilous. Yet in many organizations, there are voices, often well-intentioned, that advocate "cutting through the bureaucracy" or "throwing out the rulebook" to get quick results. The promise is usually that by discarding the cumbersome old ways, the organization can become agile and innovative overnight. It's an attractive narrative, especially in an era that celebrates disruption. But when put into practice in a complex, real-world operation, this approach often increases risk rather than reducing it.

Let's make this concrete. Imagine a new manager comes in and says, "All these approval steps are slowing us down. From now on, we'll just trust each other and move fast." They eliminate some sign-offs, skip a few reviews. For a short while, things indeed move faster. But then a problem that one of those missing reviews used to catch slips through. Perhaps a contract gets signed without legal review and later exposes the company to liability. Or a product change goes out without proper quality testing and a recall ensues. The immediate reaction is panic and blame. The organization might scramble to reintroduce the very controls that were cut, perhaps even more stringently than before, having learned a lesson. In the worst case, irreversible damage is done, a regulatory fine, a lawsuit, a hit to customer trust.

This pattern is common enough that seasoned professionals often greet "slash the red tape" ideas with a skeptical eye. They've seen episodes of naive over-correction. Processes exist for reasons, as we've noted, and removing them without understanding those reasons is courting trouble. It's akin to someone ripping out the brakes of a car to make it go faster, yes, you'll speed up initially, but the crash at the end wasn't worth it. One veteran of government reform succinctly put it: bureaucracy isn't inherently bad; it's a system built to manage risk and deliver consistency across millions of people, and ripping it apart won't magically solve problems. In other words, if you treat all those rules and checks as mere obstacles and ignore their underlying purpose, you tend to rediscover the hard way why they were there.

Even in the private sector, where "moving fast" is more culturally accepted than in government, we are seeing a reckoning. The tech industry famously championed "move fast and break things" as a mantra. But as some tech companies matured or dealt with serious consequences, that slogan lost a bit of its shine. Breaking things, it turned out, can break trust, with users, with regulators, with the public. As public reporting has noted, when Facebook's early ethos of moving fast led to real-world harms (or when various startups disrupted industries without considering safety or fairness), the cleanup was painful. The lesson many have drawn is not that innovation is bad, but that ignoring the accumulated knowledge embedded in existing systems is bad. It's fine to streamline, but not to recklessly bypass checks that are safeguarding important values.

For the readers of this book, who likely come from established organizations, this is probably intuitive. You have lived through enough "costly shortcuts" to know the pattern. Perhaps someone in your organization launched a new process or product without involving the compliance team, only to find later that they violated some rule and had to roll it back. Or a well-meaning executive declared a "flat hierarchy" to empower people, but in the absence of clear roles, decisions started falling through cracks and nothing got done. In the end, you might have had to restore some hierarchy just to get clarity again. These stories reinforce an essential point: dismissing the model wholesale is dangerous.

It's also counterproductive in terms of human factors. When leaders say, "We're just going to ignore the old way," they often demotivate or alienate the very people who know the old way best, the masters we've been talking about. Those people hear, "Everything you've been good at for years doesn't matter now." They may respond by disengaging or by actively resisting to protect the organization from what they see as a rash course. Either way, the change effort suffers. Moreover, going around established procedures usually means you're improvising new ones on the fly. Unless those improvisers have exceptional foresight (and luck), they will miss critical elements. It's like removing a piece from a machine and not realizing what function it was serving until smoke starts rising. The more responsible path, and the one we will champion in this book, is to work with the grain of the existing model, making deliberate adjustments rather than tearing things out blindly. A phrase sometimes used is "evolution, not revolution," especially when lives, livelihoods, or significant assets are at stake.

However, we must be careful here to strike a balance. Saying that dismissing the model is risky is not the same as saying the model can never change. We are not advocating complacency or a refusal to improve. The message is about respect and understanding: you have to respect what the current system does well (and why) before you try to change it. If you skip that step, you risk throwing the baby out with the bathwater. Sometimes reformers come in with an almost moralistic fervor against "bureaucracy" and, in their zeal, dismantle critical safeguards. A more nuanced approach recognizes that speed and structure need to be reconciled carefully. You want the benefits of moving faster or being more innovative, and you want to keep protection against big risks. Achieving that means identifying which parts of the process are truly value-added and which might be excess fat. If you cut, you do so with a surgeon's precision, not with a chainsaw.

Throughout this chapter, we've been building trust by affirming that the reader's current capabilities are valid and valued. This section adds an important reinforcement: not only is your mastery valued, but attempts to ignore it can backfire spectacularly. In practice, many of you have probably served as the voice of reason when some initiative threatens to bypass due process. That role, sometimes the "brake" to others' "accelerator", is often unglamorous but absolutely essential. It is much better to be the cautious driver who reaches the destination safely than the reckless one who crashes halfway. Organizations have learned this the hard way, which is why they put brakes (processes) in place in the first instance.

However, respecting the current model does not require freezing it in place. We can honor the logic and contributions of the existing system while still acknowledging that the world is changing and that the system may need to evolve. In fact, true respect for the system's purpose (managing risk and enabling complex work) might compel us to adjust it when external conditions shift. The difference is in attitude: change undertaken with respect looks very different from change undertaken with contempt. The former keeps the best of the old while carefully introducing the new; the latter risks losing wisdom and ending up with chaos. Our stance is firmly on the side of evolution with respect. As we prepare to examine where changes might be needed, we carry forward this principle that any modifications should build on the solid foundation, not tear it down recklessly.

THE SUBTLE COST OF DEEP EXPERTISE

We have celebrated deep expertise in the current model, and rightly so. But now we must turn to a more delicate issue: that very depth of expertise can also create blind spots or rigidity in perspective. This isn't a contradiction so much as a paradox of success. When someone becomes highly proficient in a stable system, they internalize its assumptions so completely that those assumptions become invisible to them. The model's logic becomes "the way the world works" in their eyes. While this enables smooth operation under known conditions, it can hinder the ability to see when conditions have changed. In short, what made a person effective yesterday may limit their ability to perceive tomorrow's needs.

It's important to approach this topic with empathy and nuance. We're not talking about willful obstinance or lack of imagination as personal failings. We're talking about a natural cognitive effect of specialization. Psychologists and educators sometimes refer to the "expert blind spot," meaning that the more expert you are in something, the harder it can be to understand what novices don't know, or to imagine entirely different ways of doing things. In an organizational sense, a veteran leader might not realize how much the environment outside the company has shifted because their mental models were formed in an earlier era that rewarded different things. They might unconsciously filter out information that doesn't fit their established framework. This is not because they are closed-minded bad actors; it's a side effect of having a strong frame of reference built over years of reinforcement. As one scholar described it, long-term accumulation of capability can produce a kind of inertia, making it hard to adapt to changes, what in her public writing Dorothy Leonard-Barton has termed a "core rigidity," the flipside of a core capability.

To make this less abstract, consider a few scenarios. A risk manager in a bank has, for decades, lived by the mantra that human judgment is the most unpredictable factor and therefore all processes must be designed to minimize discretionary decisions by individuals. This belief came from seeing cases where someone's lapse in judgment led to a financial loss or a compliance breach. Now imagine technology improves or new methods emerge that could allow more flexibility without increasing risk, perhaps some advanced system can catch errors better than manual process ever could. The deeply experienced risk manager might be skeptical or slow to trust this new approach, because it violates a core lesson they've internalized: people must be tightly overseen. Their past success in preventing incidents reinforces the idea that their caution is always justified. It may genuinely be hard for them to see a future where effective risk management could take a different form. Their caution, once entirely appropriate, could become over-caution in a new context.

Similarly, consider a project leader whose entire career has been about managing complex projects through meticulous planning, stage gates, and documentation, classic hallmarks of a waterfall methodology. They've delivered many big projects without major failure, something to be proud of. Now the industry is moving to more iterative, agile approaches. To this leader, the new approach might initially look like chaos or irresponsibility, because it lacks the familiar checkpoints they associate with safety. Their first reaction might be, "This will never work for a project of size X, because I've only seen failure when people deviated from the rigorous approach." Their wealth of experience, while valuable, can cast a long shadow, it can make it harder to embrace a new mode, even if that new mode might handle certain types of work better. The assumptions built into their mastery (for example, that change control boards are always essential) might no longer hold universally, but it takes time and sometimes a leap of faith to see that.

One might say, "Well, shouldn't the experts just stay open-minded?" Of course, openness is ideal, but we must recognize that human nature and organizational culture often conspire against it. When you are the expert, others look to you for confident decisions. Admitting that the ground might be shifting under your feet can feel like undercutting your own authority. Also, organizations have historically praised these experts for their certainty and consistency. So the system hasn't rewarded them for saying "maybe this fundamental thing I know is changing." On top of that, the expert genuinely might not perceive the small cracks that are forming in the old model, because those cracks often start as subtle signals that get drowned out by day-to-day concerns.

Another subtle cost of deep expertise is a form of tunnel vision. If you've spent decades measuring success in certain metrics (say, quarterly profit, or patient safety incidents, or product defect rates), you pay most attention to those metrics. You might miss other dimensions that are becoming important, perhaps user experience, or speed to market, or adaptability. This is not due to negligence; it's due to intense focus. The organization told you these were the critical metrics, and you delivered. But if the strategic landscape changes such that a different metric starts to matter more, experts can be caught off guard. They might even resist new metrics: "Why are we suddenly so worried about speed? If we rush, quality will suffer." That concern could be valid, or it could be a knee-jerk defense of the old priority. Deciphering which it is requires stepping back, something that is hard to do from inside the driver's seat when you've been racing down the same track for a long time.

All of this is to gently introduce the idea that evolution must build on mastery, not discard it. Why emphasize the blind spots at the end of a chapter praising mastery? Because we need to set up a constructive mindset for the rest of this book. We are not going to say that the masters of the old model were wrong or foolish. They were and are highly competent under the conditions that prevailed. But we are going to examine whether those conditions are shifting and how to adapt without losing the value of expertise. The subtle costs of deep expertise, narrowed perception, inertia, perhaps a bit of overconfidence in the eternal correctness of one's approach, are challenges we have to face with our experts, not against them. The goal is to help expertise evolve without the experts feeling that they are being declared obsolete.

This sets the stage for the pivot of the book. Up to now, our trust-building arc has aimed to say to the reader: your knowledge and skills are respected here. We are not another voice telling you that "everything you know is useless, step aside." Quite the contrary, everything you know is the foundation on which the future must be built. But, and here is the crucial but, resting on that foundation alone may not be enough if the environment around us has changed. Deep expertise can become rigid if it never adapts. The heroes of the previous model can become the skeptics or even obstacles in a new model, not out of malice but out of habituation. Recognizing that is not a condemnation; it's an invitation. It's an invitation to the masters of the current system to bring their mastery into the next era, consciously updating their mental models and shedding just the outdated parts of old assumptions.

The subtle cost of deep expertise is that it can bind us to the present (or past) just as it enables us to excel within it. The bigger the success one has achieved in a given model, the harder it is to imagine a different model. The challenge ahead, for any leader or professional reading this, is to keep the virtues of your mastery while remaining willing to question its limits. This isn't easy, but it's necessary. And it's our job in the coming chapters to guide that process in a respectful, pragmatic way.

WHAT THIS CHAPTER ESTABLISHES

We close this chapter having established a vital truth: mastery of the current organizational model is real, valuable, and still necessary. Operating within the structure we have inherited is not a trivial matter; it is a form of competence that organizations absolutely rely upon. We have seen that navigating processes and hierarchies takes skill and that such skill has been rewarded because it underpins the stability and predictability of large enterprises. We have acknowledged that the people who are fluent in this model carry much of the institution's memory and wisdom in their heads and habits. In short, nothing in this book's perspective diminishes the competence, experience, or judgment of those who make today's organizations function.

By reinforcing this, we lock in a crucial principle for any change discussions: change must build on this mastery rather than negate it. If there is one thing to carry forward, it's that evolution should feel to the master not like an attack on their expertise, but like a way to extend their expertise to meet new conditions. The first three chapters, including this one, have been about building trust and understanding the current state on its own terms. We have now completed that foundation. We have argued that the prevailing model was a rational answer to human limitations (Chapter 1), that process was a protective mechanism not mere bureaucracy (Chapter

2), and that mastery of this protective, structured model is a legitimate achievement that remains important (Chapter 3).

With this understanding in place, we can turn our attention to the tensions that are emerging. The world is not static, and even a well-mastered model can show signs of strain when conditions shift. The next step in our journey is to examine where this finely tuned system, staffed by competent, responsible people, is encountering economic and practical limits. This is not because the people became less competent overnight, but because the environment around them has changed. Coordination, which was once just a cost of doing business, may now be consuming more resources than the work itself in some areas. The pace of external change might be rendering some processes too slow, despite being expertly followed. These are the kinds of issues we will explore next.

As we proceed, remember the respect we have established. We are moving from describing why the model works to investigating where the model struggles. That transition is not a flip from praise to criticism; it is a natural progression. We will do so grounded in the belief that any solutions must incorporate the knowledge and skill of those who know the current model best. The reader should feel prepared for this exploration, having seen that we honor what exists and why it exists. Now, with that trust and mutual understanding, we can delve into the areas where even a masterfully run organization might be feeling that something isn't adding up, where coordination and process costs are rising, complexity is mushrooming, and the old trade-offs are shifting. In a sentence: "We have built a fortress to protect against human nature, but that fortress is becoming a prison for digital speed."

This next phase is not about finding fault in people or casting blame on "bureaucrats." It's about recognizing a possible inflection point in how work gets done. Because if there is strain, it implies an opportunity: perhaps new tools or approaches (emerging technologies, anyone?) could relieve some of that strain. But we will not leap there without careful analysis. We will first pinpoint what the strain is and why it's happening. Armed with that clarity, we can then discuss how to evolve the model in a deliberate, respectful way.

So, to conclude this chapter: Your current capability is valid and valued. Mastery of the existing model still matters immensely. And precisely because it matters, we owe it to ourselves to examine where that model, even in masterful hands, is becoming costly or cumbersome. Only by doing so can we ensure that any evolution is built on the strong foundation of what we know, rather than in ignorant defiance of it. The trust we've built in understanding the past and present will be our guide as we turn to the future, a future that we will approach not with disruption for its own sake, but with an aim to carry forward the best of our mastery into a new era.

4

WHEN COORDINATION COSTS MORE THAN EXECUTION

Coordination has always come at a cost. In any organization, a portion of effort goes not into making or doing things directly, but into making sure those things are done in a coordinated, consistent way. Historically, this cost was accepted as the necessary price of reliability and scale. We saw in earlier chapters that modern organizations were built around human limitations, to compensate for our inconsistency and limited span of control, we introduced roles, approvals, and processes. Aligning people and decisions across departments required meetings, communication, and oversight, and this overhead was often cheaper and safer than leaving each team to execute in isolation. Coordination wasn't a byproduct of poor design; it was a deliberate investment to prevent chaos and ensure quality. In essence, organizations agreed to "pay" a certain amount of coordination tax to achieve predictable outcomes. We can imagine a time when this tax was manageable, even minor, compared to the work of execution itself. But as organizations have evolved, something fundamental about this balance has shifted.

THE RISING COST OF COORDINATION

What exactly do we mean by "coordination" in this context? It includes all the work needed to synchronize interdependent activities, the calls and status meetings to get everyone on the same page, the email threads clarifying plans, the progress trackers and project management updates, the rounds of review and approval, and the time spent waiting on others before you can proceed. It's the energy devoted to aligning a group of people toward a common outcome. These activities don't directly produce a new product or close a sale, but they create the conditions for those things to happen without conflict or error. For a long time, such coordination work was simply part of doing business, often unnoticed. In a small company or a tight-knit team, coordination might be as simple as a quick conversation or an informal check-in. As organizations grow, however, coordination becomes more formalized: regular planning meetings, detailed documentation, cross-functional committees, multiple approval signatures, and so on. Each step is introduced for a reason, perhaps a lesson from a past mistake or a safeguard against inconsistency, and each carries its own overhead. Over years and decades, the sum of these coordination mechanisms has accumulated. We now find ourselves in an environment where the effort spent on coordination has ballooned relative to the effort spent on execution.

The data is striking. Recent published studies of knowledge workers, people whose jobs involve planning, analyzing, communicating, rather than manual labor, reveal that the majority of their time is consumed by "work about work" rather than the core work itself. Public reporting on a global survey said that around 60% of a person's working time is devoted to these coordination tasks: chasing status updates, attending meetings, writing emails and reports, managing task lists and tools. That leaves only 40% of time for the skilled, specialized work that these employees were hired to do. In concrete terms, an average employee might spend hundreds of hours per year on emails and meetings that coordinate projects, hours that used to be available for focused productive effort. Other research echoes this trend. Public reporting on Microsoft's workplace analytics, for example, suggests that knowledge workers now spend anywhere from 33% to as much as 80% of their week on coordination activities, preparing for meetings, updating plans, communicating status, rather than on direct value creation. We have effectively reached a point where many people invest more time organizing and reporting on work than actually executing it. The coordination cost has grown from a tolerable overhead into a dominant use of time.

Several factors underlie this dramatic shift. One is the increasing complexity of modern work. In earlier eras, an individual or small team might carry a task from start to finish with minimal outside input. Today, even straightforward initiatives often involve multiple stakeholders and checkpoints. We have more specialists whose work must be integrated, more compliance requirements to be checked off, more markets and regions to synchronize. In response, organizations have layered on tools and processes to manage this complexity. Ironically, some of the very tools introduced to streamline collaboration have contributed to fragmentation of work. Communication channels have proliferated, email, team chat, automated meeting transcripts, AI summaries of meetings, project management dashboards, and collaborative documents, all bringing with them constant notifications and the need to update multiple systems. It's easier than ever to send a message or schedule a meeting, so we do more of both. Each individual alignment or update is quick, even trivial, but in aggregate they create a perpetual web of "busywork." As public reporting has noted, in trying to become more connected and informed, we have accidentally "created a host of other problems that add up to a bunch of unnecessary and tedious work in order to keep track of what's going on". In short, the coordination mechanisms that served us well at one scale have multiplied and started to devour the calendar.

Another factor is that the effort required for execution, the hands-on act of doing work, has not increased nearly as much; in fact, it has often decreased. Over time, improvements in tools, technology, training, and now AI agents have made many forms of execution more efficient. Tasks that once took days of labor might now be done in hours or automated entirely. Preparing a financial report, for instance, was a manual, error-prone job decades ago; now software can compile data in seconds, with the human effort focused on review and interpretation. In manufacturing and services alike, automation and better systems have lowered the marginal cost of performing a unit of work. The result is that once a task is defined and agreed upon, actually carrying it out is often cheaper and faster than ever before. But crucially, deciding on the task and coordinating its execution have not seen comparable efficiency gains. If anything, those aspects have become more laborious because the context around each task is more complex. In the time it takes to align all the players and ensure consensus, our improved tools might have finished the work twice over. This creates a widening gap where execution is incrementally easier, but getting to the point of execution, getting approvals, clarifying scope, aligning stakeholders, is harder. The balance between the two has tilted: execution effort has dropped faster than coordination effort. What used to be the main source of cost (the sweat to build or deliver something) is now often the lesser cost, while the greater cost is the overhead of orchestrating the many moving parts involved.

Evidence of this inversion is visible in how organizations allocate their people. In published commentary, management scholar Gary Hamel observed that, since 1983, the number of managers and administrators in the U.S. workforce (the roles largely responsible for coordination and oversight) has more than doubled, even though growth in all other job categories was under 40% in the same period. In other words, organizations have added coordinating roles at a far faster rate than roles dedicated to executing work. These extra layers and staff are not inherently "bad", they often exist to handle the growing complexity of operations, but they illustrate how the weight of coordination has grown. Every additional manager, project liaison, or analyst represents an investment in aligning and supervising others. As those roles multiply, they also change how decisions are made. A traditional engineer-to-manager ratio might shift so that a greater share of employees' time (both managers and those reporting to them) is spent in communication, reporting, and cross-functional syncing. We can also see the shift in how people describe their daily jobs. It's increasingly common to hear frustration that "I spend more time in meetings than doing the actual work" or "the real work begins after 5 PM, once I'm done with all the status calls." These sentiments are backed by public survey results: large fractions of employees report that internal bureaucracy, the rules, approvals, and coordination overhead, directly slows them down. Even at companies famed for agility, the internal cost of collaboration can creep up. Public reporting has described Google, for example, finding in an employee survey that over half of its employees felt the company's bureaucracy and hierarchical processes were slowing down their work. When an organization built on "move fast" principles reaches that point, it signals that a threshold has been crossed. The cost that was once just the oil keeping the machine running smoothly has become significant enough to gum up the works.

NON-LINEAR GROWTH OF COORDINATION OVERHEAD

Part of the challenge is that coordination costs do not scale in a gentle, linear way. They tend to ramp up faster, sometimes much faster, than the growth of the organization's output or headcount. A small increase in complexity can cause a disproportionately large increase in the effort needed to coordinate. This happens for mathematical reasons. Communication in a group grows combinatorially: if you double the number of people involved in a project, you more than double the number of communication channels and possible interactions among them. With three people on a team, there are only three bilateral communication pathways; with ten people, there are 45; with fifty people, 1,225 potential paths. By the time you have 150 people interacting on related efforts, the theoretical communication pathways explode to over 11,000. Of course, not everyone needs to talk to everyone else, we put structures in place to manage communication, but that itself is coordination. We create reporting lines, liaison roles, standardized meeting cadences, all to manage this exponential complexity. Each added node (whether it's a person, a team, or an additional stakeholder group) introduces new possible points of misalignment that must be tended to. Adding just one more stakeholder to an approval chain, for instance, doesn't just add a single step of work; it introduces the possibility of new back-and-forth discussions, revisions, and delays that compound geometrically as the interactions multiply. If three departments must sign off on a proposal instead of two, suddenly there are more permutations of discussions, Department A may need to clarify something with C that didn't matter when it was just A and B, and so on. The alignment effort multiplies in ways that aren't immediately obvious when looking at an org chart or project plan.

This compounding nature is especially evident in projects that span multiple functions or regions. Consider a cross-functional initiative, say launching a new product that involves the design, engineering, marketing, legal, and operations teams. The execution work, designing the product, coding the features, setting up manufacturing and distribution, might be well-understood within each team. But to bring it all together, those teams must constantly coordinate: marketing needs to know the features to craft messaging, legal needs to review the plan for compliance, operations needs forecasts from marketing for production, and engineering needs feedback from all sides. If any one of these threads gets held up, others may have to pause. Now add a global dimension: the same product launch coordinated across multiple countries, each with their own regulatory reviews and market considerations. A straightforward task transforms into a complex program of meetings across time zones, version control on documents, sequential approvals from different corporate offices, and frequent recalibrations when one piece of the puzzle changes. What might begin as a two-month engineering timeline can stretch into half a year of overall project time, with the lag mostly attributable to coordination: ensuring everyone has agreed and everything is lined up. The more interdependencies and checkpoints in the effort, the more the cost curve of coordination shoots upward. As one management article succinctly noted, "doubling an organization's headcount more than doubles coordination overhead because interaction possibilities expand combinatorially". This is a structural reality, not a mere matter of individual efficiency. Even highly skilled, well-intentioned people can find themselves entangled in a web of necessary consultations and sign-offs that grows denser with each additional link in the chain.

Crucially, coordination costs don't just grow with size; they also grow with complexity. An organization might remain the same size but take on more diverse projects, more product variations, or more stringent compliance regimes. Each new dependency or requirement can introduce what you might call a tax on progress, an extra layer of alignment needed. For example, a new regulatory rule might require that two additional departments review every marketing campaign. Suddenly, a task that used to involve three people now involves five, with all the attendant scheduling and consensus-building that entails. The cost per campaign hasn't increased in the creative work, but it has in the coordination work. Or imagine a company that introduces a matrix structure where employees report to both a functional manager and a project manager. Such a structure can have great benefits for information sharing, but it also means that any decision might need to be communicated and vetted through two chains of command instead of one. Again, the effort to align increases faster than the work itself. These examples underscore a sobering point: as organizations expand in scope or ambition, the coordination overhead can surge out of proportion, quietly turning into a dominant cost driver. It's not that anyone chooses this outcome; it emerges from the mathematics of collaboration and the accumulation of safeguards and linkages we've put in place.

WHEN COORDINATION BECOMES THE BOTTLENECK

A telling sign that coordination has overtaken execution is when the pace of work is no longer set by how quickly teams can carry out tasks, but by how quickly they can get aligned and obtain approvals. In effect, work starts to move at the speed of consensus. Employees begin to sense that waiting for decisions or clearance is a bigger part of their job than doing the work itself. Have you ever completed a piece of work in a day, only to spend the next two weeks getting it reviewed, signed off, and synchronized with others' schedules? When stories like that become common inside an organization, the system is straining under coordination load. Even small changes or routine updates end up queuing behind layers of alignment steps and gating processes. It's not because people are dragging their feet; often everyone involved is working diligently. It's simply that the system's design requires multiple sequential handshakes for anything to happen, and those handshakes take time. Each one might be reasonable, a review to ensure quality here, a checkpoint to avoid redundant work there, but in sum they form a critical path far longer than the work itself. A simple decision that once could be made on the spot now might require a committee meeting next week because so many stakeholders are involved. A minor feature tweak might sit in "approval limbo" until various managers and teams have weighed in, even though the engineer could implement it in an afternoon once the go-ahead comes. Over time, the default answer to "When can this be done?" shifts from "as soon as we finish the task" to "as soon as everyone concerned is looped in and agrees." In other words, execution is waiting on coordination.

From the outside, this slowing of output can be perplexing, resources are at an all-time high, teams are packed with talented people, and yet progress feels sluggish. Inside the organization, the people directly responsible for deliverables often feel it most acutely. They notice that projects that used to turn around in a month now take a quarter, not because the work is harder, but because there are twice as many update meetings and documentation requirements. A software developer might recall the "good old days" when she could focus on coding, whereas now half her week is absorbed by cross-team design reviews and status syncs. A salesperson may find that closing a deal involves navigating an elaborate internal approval workflow for discounts and custom terms, where previously a quick manager sign-off sufficed. These are the human faces of coordination overhead, talented professionals feeling a drag on their ability to get things done. They experience more drag, more rework, and more idle time waiting for others, even if on paper the company's processes are supposed to be efficient. In many cases, these front-line practitioners see the strain early. They might start voicing concerns: "It feels like I spend more time justifying the work than doing it," or "By the time all the reviews are done, the opportunity has passed." These comments aren't excuses; they are important signals from those who deliver outcomes that the model is overextended.

It's important to recognize that this slowdown is a systemic issue, not a sudden loss of will or capability. Often, senior leadership doesn't immediately see the same degree of pain because the high-level metrics might still look acceptable, deadlines are eventually met (thanks to Herculean effort and overtime), and deliverables eventually go out (after navigating the gauntlet). If anything slips, it may be attributed to normal growing pains or temporary resource constraints. There can even be a misinterpretation: upper layers might perceive the frustrations below as resistance to process or a need for better discipline, when in fact those doing the work are already extremely disciplined, they're just bogged down by too many channels of coordination. This disconnect can lead to a tragic cycle in some organizations: managers implement even more reviews and controls to "get a handle" on delays, not realizing that those interventions further increase coordination load. The people closest to the work feel the squeeze tightening, while those farther up see only a complex situation that seems to require more oversight. The result is a feeling of work life slowing to a crawl despite everyone's best intentions. In surveys, bureaucracy consistently ranks as a top frustration. Global benchmarks show that bureaucracy (defined by employees as slow or complex processes and "red tape") garners the lowest favorability scores in workplace experience, with less than half of employees viewing their organization's bureaucracy positively. In other words, a majority feel trapped in slow, inefficient procedures. It's telling that even managers and executives often rate the burden of internal process worse than front-line employees do, a sign that the drag of coordination is palpable at all levels. When nearly everyone in an organization senses things are too slow, it's usually because the predominant activity keeping them busy is not creating value but negotiating the maze that's been built to protect value.

What does it look like, day-to-day, when coordination becomes the main event? It can be subtle at first. You'll see calendars filled with back-to-back meetings, leaving little time to do concentrated work. You'll hear people saying they need to "find some time this weekend to catch up on real work." Projects will have plenty of activity, emails flying around, frequent check-ins, but those activities increasingly revolve around updates on status rather than leaps of progress. People start creating elaborate dashboards and reports to monitor the very complexity we've created, which itself becomes another layer of work. In effect, the organization starts expending significant effort to manage itself. Coordination becomes the work. This is not a dramatic proclamation but a gradual shift in emphasis. At a certain tipping point, the majority of intellectual effort, creative energy, and working hours are spent on maintaining alignment: communicating, tracking, reviewing, and adjusting, rather than on producing new value. A financial analyst might spend more time updating the many spreadsheets that feed into leadership reports than analyzing market data for insights. A product manager might devote more effort to getting buy-in from all departments on a minor feature change than on actually defining the feature. These are signs that coordination has crossed from being a supporting function of work to being the central function. Meetings, updates, approvals, and reviews start to consume more time and energy than producing the outcome itself. In such environments, employees can understandably feel like hamsters on a wheel, constantly in motion, but not necessarily moving forward as fast as they could. Importantly, this is usually a slow progression, not an overnight switch. An organization doesn't wake up one morning to find, suddenly, that coordination is eating 60% of everyone's day; it creeps up as teams grow, as new control steps are added one by one, as well-intentioned processes accumulate. Because it's gradual, it's easy to miss the moment where the balance tips and to rationalize each new bit of overhead as "just part of how we have to operate now." But when coordination becomes the dominant consumer of resources, the organization's whole dynamic changes. We see diminishing returns on new initiatives (so much effort, so little output), and often a kind of quiet demoralization sets in, as people feel they are pouring their time into a system rather than into their craft.

To be clear, none of this is about casting blame on individuals. The engineers in those meetings are not lazy for not coding more, they're following the required process. The managers asking for more reports aren't trying to create busywork, they genuinely need visibility to feel confident in decisions. And the leaders who instituted multiple approval gates didn't do so arbitrarily, they often reacted to prior issues or risks that those gates are meant to catch. In fact, it bears emphasizing that the existence of high coordination costs in an organization is usually not due to incompetence or malice. It's a byproduct of success and rational management choices. As we discussed, coordination was traditionally seen as a necessary expense to ensure consistency and control in the face of human limits. Those guiding principles haven't changed. What has changed is the context: the economics of how much effort goes into coordination versus execution. In a sense, the "model" of a well-run organization, with its planning cycles, cross-checks, and managerial oversight, is being asked to operate at a scale and speed where its cost structure no longer balances out as it used to. The model isn't suddenly broken; it's just stretched. The costs that were once acceptable and absorbed in the background have become so large that they command attention. This is fundamentally a problem of economic balance, not of individual willpower or talent. When coordination starts costing more than execution, it signifies an imbalance between two categories of work that should ideally complement each other. It's like a ship whose ballast has shifted, not because the crew failed at their jobs, but because the cargo load changed.

Researchers sometimes frame this as a shift in what the organization is optimizing for. When coordination costs dominate, an organization can unconsciously start optimizing for internal alignment at the expense of output. You see extremely well-documented decisions, impeccably vetted plans, thoroughly briefed teams, and perhaps slower delivery or missed market opportunities as the trade-off. It's not that anyone wants to prioritize paperwork over results; it's that the system's internal logic begins to reward the things that ensure alignment (since misalignment is painful) more than the things that create external impact. Everyone feels the strain: the people on the ground feel slowed and sometimes frustrated, and leaders feel perplexed by why more resources aren't translating to proportional gains. But it's critical to recognize that this outcome is not due to laziness or poor leadership. On the contrary, it often arises because people are diligently following the rules and processes that were put in place for good reason. It's a case of doing "everything right" according to the old playbook, and yet finding that the results are underwhelming.

Why isn't it someone's "fault"? Because the organization as a whole is operating exactly as it was designed to. The design just happened to reach a point where its own maintenance work (coordination) outweighs its productive work. As one organizational psychologist put it, bureaucracy and coordination mechanisms are not inherently bad, they bring order, consistency, fairness, and safety to operations. Those benefits are real. A bank, for instance, has heavy coordination overhead in the form of compliance checks and internal audits, and for good reason, as they prevent costly errors or fraud. The trouble emerges when those mechanisms grow to a point that they begin to slow people down or erode trust in the system's efficacy. A process that once added assurance now adds anxiety because it's seen as getting in the way of serving the customer or achieving the mission. Yet even that isn't usually because of "bad" decisions; it's often a result of applying yesterday's successful control strategies to today's larger and faster-moving environment. The intent, to maintain control and quality, is valid. The outcome, however, is that the balance of effort has flipped. The economic equation that justified coordination (i.e. coordination cost < cost of errors or inefficiency without it) may no longer hold. Instead, coordination costs have compounded to become the primary cost. At that point, the organization is essentially paying more in oversight and alignment than it would likely lose if it trusted people a bit more and coordinated a bit less. That is a precarious situation: it means the organization is, in effect, taxing itself so heavily that the tax exceeds the value of what it's protecting.

Recognizing this inflection point is not about fault-finding; it's about seeing a structural pattern for what it is. The system is straining under an out-of-date economic balance. What worked at a smaller scale or with more expensive execution doesn't automatically work when coordination is the bottleneck. This is why the experience of work slowing down is often a symptom of a deeper shift. It signals that we have crossed from one regime to another, from a regime where adding coordination improved outcomes, into one where adding coordination yields diminishing or negative returns. It bears repeating: this is not because people failed. It's because all the rational, well-meaning additions of process, oversight, and alignment, accumulated over time, have tipped the scales. It's an unintended consequence of trying to do the right thing by the old measures. In many ways, it's a testament to the prior success of the model that we only notice the strain once very high levels of coordination are in play.

So, what happens next? Often, organizations confronted with these symptoms react in a predictable way: they seek to restore a sense of control and coherence by doubling down on what has worked in the past. If projects are slowing due to coordination issues, the instinctive response might be to introduce even more coordination, more status checks, more detailed project plans, additional oversight committees, in hopes of catching problems early and enforcing alignment. Paradoxically, this can create a vicious cycle, adding further to the coordination burden. It's a bit like trying to solve a traffic jam by adding more traffic cops to direct cars: beyond a point, the very act of directing becomes part of the congestion. When coordination costs have already become dominant, piling on extra reviews and reporting may reassure leadership that everything is being seen and managed, but it doesn't necessarily speed things up. In fact, it often adds another layer of work for teams to perform (preparing reports, attending extra meetings) without addressing the root cause. This dynamic, organizations responding to coordination strain with even more control mechanisms, is a common pattern. It comes from a good place: leaders want to ensure nothing critical is slipping through the cracks as things get more complex. However, it can inadvertently create what might be called a consensus theater, a flurry of oversight activities that give the appearance of control, while actual execution grinds along slowly.

We will explore that phenomenon in subsequent chapters. For now, the key takeaway is the foundation we've laid: coordination effort has grown massively relative to execution effort, to the point that it can overtake it; once coordination becomes the dominant cost, the organization's engine begins to lose efficiency and strain under its own weight; the need for constant alignment, review, and dependency management creates compounding overhead that slows work down; and this slowdown, felt by so many in their daily work, is a systemic effect of that shifting balance, not a failure of the people or an indication that the principles of leadership and structure that served us well are suddenly wrong, but a sign that their economic footing has changed. The model of organizing human work that we inherited still works, but less cleanly than it once did. Its cost structure has quietly transformed. Recognizing this is the first step toward understanding how we might rebalance our organizations for a new era. Before we consider any changes or solutions, we must fully acknowledge this reality: when coordination costs more than execution, the old assumptions about how to run an efficient organization no longer hold the way they used to. In the next chapters, we'll examine how organizations typically react to this imbalance and what that means for the way we exercise control and make decisions. For now, it's enough to observe that a threshold has been crossed. Coordination, the very mechanism that once enabled success, has in many places become the heaviest drag on success. The challenge ahead is not about blaming bureaucracy or throwing out all process, it's about carefully restoring the balance so that alignment serves the mission, rather than inadvertently subverting it.

WHAT THIS CHAPTER ESTABLISHES

This chapter establishes that coordination is no longer a background cost of organizational life, but in many environments has become the dominant consumer of time, attention, and energy. The core insight is not that coordination is unnecessary or poorly designed, but that its economic balance has shifted. What once served as a rational investment in reliability now, at sufficient scale and complexity, begins to overwhelm the very work it was meant to support.

The chapter makes clear that this shift is structural, not behavioral. People are not coordinating more because they are inefficient, resistant, or poorly managed. They are coordinating more because the organization is operating exactly as it was designed to operate under conditions that no longer hold. The accumulation of reviews, handoffs, approvals, and synchronization mechanisms is the predictable outcome of applying a human-centric control model to an environment where execution has become faster, cheaper, and more automated.

By showing how coordination costs grow non-linearly with size and complexity, the chapter explains why organizations often feel simultaneously well-resourced and slow-moving. It reframes familiar frustrations - endless meetings, approval bottlenecks, work happening "after hours," and progress that lags effort - as symptoms of an economic inversion rather than management failure. When coordination overtakes execution, the organization begins optimizing for alignment rather than impact, not by choice but by necessity.

This chapter sets the conditions for what follows. It demonstrates that once coordination becomes the bottleneck, attempts to restore performance through additional oversight and synchronization will only deepen the problem. The system responds to strain by amplifying the very mechanisms that caused it. This creates the opening for the next chapters, which examine how organizations attempt to reclaim control under these conditions, why those attempts often turn performative, and why restoring balance requires a fundamental shift in how control is expressed rather than simply more of it.

5

THE ILLUSION OF CONTROL

As organizations grow and projects become more complex, a pattern reliably emerges: the instinctive response to rising coordination strain is to add more control. In a small team, alignment can be achieved with a quick conversation; in a larger enterprise, that same alignment takes structured reviews, approval workflows, and formal status reports. With every new interdependency or missed signal, leadership's natural impulse is to introduce another checkpoint or committee to prevent surprises. These additional reviews, approvals, and governance forums are usually introduced not because managers crave power, but because they crave confidence and predictability in outcomes. In other words, each new control mechanism is a rational safeguard against uncertainty. It is a protective response: when execution starts to slow or wobble under the weight of coordination, doubling down on oversight feels like the responsible thing to do.

This reaction has deep roots in the way modern organizations were designed. As discussed earlier in this book, many structures - escalation paths, approval chains, reporting lines - originated as sensible answers to the limits of human attention and consistency. If one person might overlook a risk, a second pair of eyes should review it. If project teams might drift apart on priorities, weekly steering meetings will keep them in sync. Over time, these layers accumulate. Whenever a mistake or miscommunication occurs, the follow-up question is almost always: "How do we prevent this from happening again?" The immediate solution tends to be,

"Add a process or checkpoint so it doesn't slip through next time." Each fix is reasonable in isolation. Indeed, more control often works in the short term: errors are caught, misalignments are flagged, and leaders feel more at ease knowing a new safety net is in place.

However, something insidious happens as these nets multiply. The end state is an organization encrusted with procedures that all ostensibly serve to keep work on track. There are approval matrices for expenditures, committee charters for product decisions, sign-off gates at every project phase. All of it creates an aura of rigor and thoroughness. Yet the very abundance of oversight that promises to make coordination easier can start to have the opposite effect. Each added layer carries a cost in time and complexity. Decisions that once took an afternoon now stretch over weeks, awaiting slots on busy agendas and juggling multiple approvals. Public reporting on a recent study at Microsoft found that knowledge workers spend huge portions of their week on coordination rather than actual execution. By one measure, over 33% and in some cases up to 80% of their time is devoted to meetings, status updates, and other coordination tasks. Such statistics underscore the paradox: the mechanisms meant to help control the work can end up dominating the work.

It's important to note that none of this expansion of control is born from bad intentions. On the contrary, most of these processes are put in place by conscientious leaders and teams trying to be responsible. As alignment efforts increase and execution slows, they seek reassurance through additional control. In their shoes, many of us would do the same. If a major deadline was missed last quarter, it seems only prudent to require more frequent progress reports. If a product launch stumbled due to a legal oversight, the obvious remedy is to insert a legal review step in the pipeline. These reactions treat control as a form of insurance. The organization essentially says: "We'll feel safer if we can see and approve everything." Control has become the default response to complexity, a kind of reflexive antidote to the anxiety that something might slip through the cracks.

Yet reassurance and real control are not the same thing. An organization can easily become over-controlled in structure while under-controlled in practice. To see why, we must examine a pair of pervasive illusions that accompany this proliferation of oversight. The first is that if leaders and stakeholders can see more of what's happening, they must necessarily be in control of it. The second illusion is that if everyone is busy following the processes, the organization must be making progress. Both assumptions turn out to be dangerously seductive and often false.

VISIBILITY IS MISTAKEN FOR CONTROL

Modern management culture puts a premium on visibility. We equate being well-informed with being in command of the situation. It's no surprise, then, that as coordination challenges mount, organizations respond by flooding themselves with information about the work. Dashboards are implemented to show real-time metrics. Dozens of metrics are tracked and reported upward. Managers demand frequent status emails and fill calendars with update meetings. The logic seems sound: if we see the work, we can control it. If every risk and delay is surfaced on a colorful chart, nothing will catch us off guard. Unfortunately, this conflation of visibility with control is largely an illusion.

Consider the case of a classic micromanager. Such a manager might insist on daily check-ins, detailed progress reports, and being copied on every email. From their perspective, all this reporting creates a comforting sense of oversight. They feel in control precisely because they have constant visibility into trivial details. A common observation in public management commentary is that some leaders "mistake visibility for control, demanding constant proof of progress". The team is indeed spending a lot of time producing visibility via slides, updates, and data, but that effort does not inherently improve outcomes. In fact, the more time spent on reporting, the less time is available to actually do the work. The manager's presence looms over every decision, yet decisions do not necessarily get better; they just get slower. Awareness has been achieved, but influence over results has not. It's the difference between knowing every move your players make in a game versus coaching them effectively to win.

This pattern plays out on a larger scale across organizations. Today it's common for senior leaders to have enterprise-wide dashboards and detailed reports on dozens of initiatives. Calendars are full of review meetings where slide decks of metrics are presented. From the boardroom's perspective, all lights might appear "green." The dashboard glows reassuringly with traffic-light indicators and burndown charts. There is motion everywhere: teams diligently updating entries in project management tools, managers circulating reports. This frenetic activity produces an image of control and diligence. It gives people the comforting sense that something is being done, as one author put it. After all, if we can measure it and monitor it, how hard can it be to manage it?

The reality, however, is that awareness is not the same as control. A dashboard can tell you how many tasks are delayed, but it can't tell you why or what to do about them. A weekly status meeting can surface a problem, but it doesn't automatically resolve it. The organization can become exceedingly good at exposing every issue in real-time and still find itself helpless to actually change course. In fact, there is a perverse scenario that many professionals will recognize: as projects falter, the reporting requirements tend to increase. Executives request more data, more frequent updates, tighter monitoring. All this visibility can create a flurry of involvement that masks a declining ability to actually steer outcomes. Leaders feel engaged and informed, their inbox fills with detailed reports, yet the underlying problems persist unresolved. The signals are all there, but converting signals into effective action is another matter entirely.

One reason visibility seduces us is psychological. When we're distant from the front lines, seeing information provides a sense of participation. A senior leader, far removed from day-to-day work, may not directly influence a project's course, but by scrutinizing metrics they feel as if they are contributing. It's an understandable human bias: if I'm looking at it, I must be controlling it. Furthermore, an obsession with measurable indicators can become its own trap. We tend to value what we can measure and report; if something is on a graph, it feels real, whereas intangible factors (like team morale or implicit knowledge) feel "unmanageable". So organizations optimize for the seen: they pour energy into visible outputs, reportable milestones, numerical KPIs, all the artifacts that can be captured in a dashboard. But progress, as Peter Sowrey has publicly observed, often "does not live there." True progress might be a quiet change in behavior or a decision that removes an obstacle, things that don't show up on standard reports.

In the end, visibility is just that: seeing. It adds surface area without necessarily reducing underlying complexity. Knowledge about every moving part of a project does not automatically grant the leverage to direct those parts. It can even give a false sense of security. Leaders might think, "Nothing can go wrong because we're watching everything." But as we'll explore, this confidence can be dangerously misleading. Being inundated with data can numb an organization to action, everyone assumes that because the issue is known and logged, someone somewhere must be handling it. The comfort of transparency can lull teams into believing outcomes are under control when in fact the ship is merely drifting with all instruments brightly lit.

DECISIONS MULTIPLY AS RISK IS FRAGMENTED

One side effect of trying to control complex work is the fragmentation of decisions. As organizations add specialized roles and oversight functions, a single stream of work can splinter into dozens of micro-decisions, each handled by a different party. This often originates from a sensible place: spreading risk and responsibility so no one person bears the full brunt of a mistake. If launching a new product involves technical risk, legal risk, financial risk, and brand risk, an organization will naturally assign each of those to different experts. In theory, this should improve the quality of decisions, each aspect gets scrutiny from someone qualified. In practice, it can produce an explosion in the number of decisions required and diffuse the overall accountability for the final outcome.

Think of a scenario in which an initiative that once required a single strong decision by a general manager now requires approval by a committee of five, plus sign-offs from legal, compliance, finance, and IT. Work that could have been greenlit in one meeting now passes through a gauntlet of checkpoints. Each checkpoint reduces individual exposure ("at least my function won't be blamed if something goes wrong in that area") but increases the total volume of decision-making overhead dramatically. What used to be one decision is now ten. Every department head might add a few conditions or requests, fragmenting the original decision into a series of smaller ones.

This phenomenon is structural, not merely the result of cautious personalities. As discussed in Chapter 1, modern enterprises distribute expertise and oversight for good reason, no single person can master all domains. But the cost of this distribution is that responsibility becomes sliced into thinner and thinner pieces. Decisions proliferate because each risk domain (budget, safety, quality, security, etc.) installs its own micro-decisions to stay "in control" of its slice. It's a bit like an assembly line that must stop at every station for inspection before moving on. The product will get many eyes on it, but the journey might be slow and disjointed.

One outcome is decision latency: even trivial choices take a long time when many stakeholders must weigh in. A simple change that a frontline employee could have made in minutes might spend weeks in an approval queue bouncing between departments. Another outcome is what we might call decision dilution. By the time a course of action has navigated all the sign-offs, it may bear little resemblance to the decisive vision that might have existed at the start. Each approver optimizes for their concern (legal minimizes risk exposure, finance minimizes cost, etc.), and the resulting compromise may be suboptimal for the whole. It's as if responsibility for the final outcome has been broken into fragments, with nobody owning the integrated result.

Crucially, more decisions do not necessarily equal better decisions. When responsibility is so fragmented, it can happen that no one feels empowered (or obligated) to judge the overall picture. Each actor might fulfill their narrow role diligently, yet the composite outcome can still be poor. Large project failures often reveal this pattern. In post-mortems, you hear things like: "Well, the engineering team signed off on the technical components, the risk committee approved the risk mitigation plan, the finance team okayed the budget... yet somehow the project as a whole was doomed from the start." Everyone was responsible for something, but no one was responsible for the thing. In such an environment, the organization can suffer from a sort of accountability vacuum in the center of an abundance of process.

It's important to emphasize that this multiplication of decision points isn't about incompetence or malice. It results from logical choices to distribute risk. Each additional sign-off exists because, historically, leaving that aspect to one person or team caused pain. Perhaps a contract was signed that Legal hadn't reviewed, leading to a fiasco, henceforth Legal must approve all contracts. Perhaps an unethical practice slipped by unnoticed, hence a new ethics committee must clear certain decisions. Each addition is defensible. But over time, the organization ends up with so many guardians at the gate that progress stalls. And ironically, while all those guardians reduce certain risks (the ones in their silo), the overall risk can increase because the organization becomes less agile and issues fall between silos. When ten people each own one-tenth of a decision, who is accountable for the outcome as a whole?

Thus, as risk is fragmented and spread thinly across roles, the organization faces a new kind of hazard: too many decision-makers, and too few decision owners. The total number of "decisions" being made skyrockets (each with its paperwork and meetings), giving an appearance of thoroughness. But this can mask the lack of decisive leadership on the issue. An observer might see a flurry of decision activity and assume all angles are covered. Yet because responsibility is so diluted, critical integrative judgments - those that consider the trade-offs and the big picture - may not be made at all. It's a classic case of missing the forest for the trees: every tree is being carefully inspected, but the forest is left untended.

This sets the stage for the phenomenon at the heart of this chapter. With so many controls, so much reporting, and so many cooks in the kitchen, organizations often slip into a mode of operation where appearing to decide and control things becomes a substitute for actually doing so. The process of decision-making turns performative. Meetings and reviews become rituals where everyone plays their part, risk is formally "managed" in slices, and all the visible signs of diligence are on display, but the substance of decisions, the real follow-through and accountability, evaporates. This is the essence of consensus theater.

THE RISE OF PERFORMATIVE DECISION-MAKING

In theater, actors go through motions scripted to give the audience a certain impression. In organizations, a similar dynamic can take hold: decision-related rituals that give the impression of diligent decision-making, without the real impact. This is what we call consensus theater, the elaborate staging of decisions, as opposed to actual decisions that drive action. It arises subtly in environments where coordination costs are high and accountability is diffuse. Over time, people learn that showing up and playing the role of decision-maker is often safer than actually making a bold decision.

What does consensus theater look like in practice? Imagine a recurring meeting labeled an "approval committee" or "steering group." Participants gather, a presentation is given reviewing all the analysis and options, and after much discussion there is a nominal agreement on a course of action. Everyone in the room nods, yes, we will do X. But then the meeting ends, and nothing really happens. Weeks later, the same issue might resurface in another forum, perhaps slightly rephrased, and the whole discussion plays out again. The same issues get discussed repeatedly without resolution, decisions are nominally made but never executed, and stakeholders later express surprise that the outcome they thought was decided hasn't materialized. In consensus theater, the performance of deciding has occurred, yet the organization remains stuck in the status quo.

A few telltale symptoms often signal you're witnessing consensus theater. One is a proliferation of meetings where alignment is performed more than achieved. People go around the table to state their concurrence or minor objections; minutes are meticulously recorded; perhaps a document is signed off. But afterward, team members privately confess uncertainty about who is actually going to do what. Another symptom is decisions that are reversed or ignored with little consequence. If a decision truly has an owner, you'd expect follow-through. But in consensus theater, a decision can be "made" in one meeting and then quietly undermined or rediscussed in the "meeting after the meeting." It's not that anyone overtly defies the decision; it's that the decision never had teeth to begin with. It was a script everyone read from, then they went back to their desks and continued as before.

When everyone is expected to agree, teams often slide into a polite sham of agreement. As one leadership coach described, "nobody wants to be the dissenter. So they smile, nod ... and quietly keep doing things their way." In those cases, "you don't have alignment, you have consensus theater." The group may appear aligned in the room, but the real decisions are happening elsewhere (or not at all). The rituals of decision-making, presentations, approvals, signatures, become more about showing diligence than achieving outcomes. Documents are produced "for the record," not because they will change anyone's mind. Approval signatures are collected as a way to say "we consulted everyone" rather than to truly empower everyone to act.

It's important not to misinterpret this as conscious cynicism by individuals. Usually, no one sets out to create theater. Rather, it emerges as an unintended consequence of the structures we described earlier. When responsibility is blurred and no one person can easily make a call, the organization gravitates toward collective rituals as a substitute. People with experience in such systems learn how to navigate these rituals. They know, for instance, that bringing up a controversial point in the formal meeting might be futile, so they negotiate in backchannels later. They know that getting the paperwork "checked off" is often more important for appearances than what the paperwork contains. These are coping behaviors in a system where direct, accountable decision-making has become difficult. In Chapter 3, we discussed how seasoned professionals develop mastery in their domains; here, unfortunately, even the most well-meaning and skilled individuals are compelled to participate in the pageantry because that's how the system operates. In a high-coordination-cost environment, performative decision-making can feel like the path of least resistance. It's less risky to follow the script, hold the meeting, sign the form, report the consensus, than to unilaterally drive a decision that might step on someone's toes.

Consensus theater, at its core, is a collective comfort ritual. It signals to everyone that due process was followed. All the actors, the approvers, the reviewers, the chairs, have played their parts and can assure their constituents that diligence was observed. From a distance, it looks like a fully functional decision-making process: so many meetings, so many documents, surely this organization is rigorous and well-run! But this is a facade, an illusion of control. The flurry of decision activity is often inversely proportional to actual decisive action. As one business writer observed, many companies "create enormous movement without altering the underlying physics of how decisions are made. The system stays the same, only busier". In consensus theater, the busyness of decisions creates a false confidence that progress is being made, when in fact the organization might be circling in a holding pattern.

The rise of consensus theater is understandable in human terms. When real control feels out of reach, when problems are so complex or cross-functional that no single person can grab the wheel, people resort to performing control. It's analogous to cargo-cult behavior: mimic the form of successful decision-making in hopes of reaping the same results. You gather everyone, review every detail, and pronounce a decision, because that's what a well-run organization should do. But without clear ownership and accountability (things we'll return to shortly), those decisions don't stick. They lack the force to overcome inertia.

WHY THIS FEELS SAFER THAN IT IS

If consensus theater is so ineffective, why do smart, responsible professionals tolerate it or even embrace it? The answer lies in how it feels. In the moment, a performative decision process reduces anxiety for everyone involved. For leaders, having a structured process with many participants provides reassurance: multiple stakeholders were involved, numerous eyes vetted the plan, and formal sign-offs were obtained. This visible action and collective involvement spreads out the perceived risk. It feels safer because any potential blame for a bad outcome would be diffused among all the signatories and committee members. There is comfort in the thought that "we all agreed on this course, so no single person is solely at fault if it fails."

Formal approvals and sign-off rituals also create a sense of diligence. People think, "We've followed all the steps, checked all the boxes, what more could we have done?" It's psychologically soothing to believe that process equals protection. The more elaborate the decision protocol, the more it can seem like a bulwark against chaos. This is akin to the security rituals seen in airports or corporate IT, sometimes derided as "security theater", where extensive visible measures make everyone feel safer even if some measures have little real effect. In organizations, consensus theater similarly gives a false sense of safety. Leaders sleep better knowing that a major proposal went through five committees and a board review. Employees feel less exposed if they got sign-off from their boss and peers before executing an idea. Everyone is somewhat insulated by the layers of collective decision-making.

The social aspect cannot be ignored either. Being part of big decision meetings can make participants feel important and involved. It scratches an itch for inclusion and transparency: "nothing was decided in secret, we were all in the loop." From a trust and culture standpoint, this openness is often well-intentioned. Many organizations earnestly believe that involving more people in decisions will lead to better buy-in and reduce the chance of missing something critical. And to some extent, it does improve buy-in, people are less likely to later object to a decision they were party to. So consensus theater does serve a purpose: it preemptively spreads consensus, or at least the appearance of it, which can stave off open conflict. It creates a false harmony that leaders might mistake for true alignment.

However, the safety that consensus theater provides is largely illusory. The comfort is superficial, and it comes at a cost. All those layers of process undeniably slow down decision-making dramatically. The organization becomes sluggish, reacting to challenges with delayed, diluted actions. In a fast-moving environment, slowness is risk. By the time a heavily vetted decision emerges, the situation on the ground may have changed, or an opportunity may have passed. Moreover, the substance of decisions often fails to improve despite the rigor of the process. Sitting through a two-hour committee debate does not guarantee that the decision reached is any wiser than what a focused small team would have decided in 30 minutes. In fact, lengthy group processes sometimes produce worse outcomes, e.g., overly conservative choices that represent the lowest common denominator of a large group (the proverbial design-by-committee camel). So while everyone feels they did their due diligence, the organization may end up choosing a path that pleases everyone a little but solves nothing decisively.

There's also the danger of complacency. When people rely on the existence of a process to catch errors, they may individually become less vigilant. Psychologists have observed this in safety contexts: visible protocols can lull workers into letting down their guard, believing the system will keep them safe. The same applies here. If an initiative fails despite going through the full consensus theater, participants might shrug and say, "Well, we followed the procedure; it was just bad luck." The ritual provided cover, a sense that no further personal responsibility was needed. But what if the procedure itself was flawed? What if the risk wasn't in the known factors the reviews focused on, but in the unknown factors that no one took ownership of? Consensus theater can blind an organization to such possibilities. Everyone assumes that with so many people involved, someone will notice any critical issue. It creates a lattice of assumed vigilance where in reality there may be gaps. As safety experts warn in public reporting, a false sense of security can be more dangerous than no security at all, because it reduces the urgency to actively seek out and address threats.

Additionally, consensus theater can breed a subtle form of risk inflation. Because the process to make any change is so onerous, organizations tend to bundle decisions into larger, infrequent batches (since getting approvals is hard, you pile multiple changes into one big decision). This ironically can increase risk: instead of iterating in small steps where mistakes are minor, the organization lurches in large moves approved in grand summits. By the time the next decision window opens, problems may have grown. And since everyone feels shielded by the group, there might be less proactive course-correction in between. The very structures aimed at reducing risk (lots of oversight, big meetings) can result in slower recognition of issues and slower responses, quietly increasing the overall risk exposure.

Consensus theater feels safe because it offers psychological and political cover. It scratches the itch for control in uncertain situations by showing us tangible activity - meetings, reports, signatures - that we equate with managing risk. But this feeling of safety is an illusion. It is safety in the way that a child feels safe wearing a cape because it worked for their superhero e.g. comforting, but not causally effective. The organization might feel it has done everything possible, when in truth it has mainly exhausted itself with ceremony. The real dangers lie in what was not addressed: the speed of decision-making, the quality of judgment, and the clarity of accountability. Those are the factors that actually keep an organization safe and effective, and they can be eroded by too much reliance on process-as-protection.

HOW CONSENSUS THEATER DISTORTS ACCOUNTABILITY

Perhaps the most pernicious impact of consensus theater is the way it distorts accountability. Accountability is the backbone of execution: it's the answer to the question "Who is responsible for this outcome?" In a healthy decision-making environment, accountability is clear. Even if many people contribute to a decision, there is a designated owner who will take charge of implementation and bear the consequences of success or failure. Consensus theater muddies all of this. When every decision is the product of many small approvals and inputs, pinning down who owns the result becomes extremely difficult.

In the thick of performative decision-making, one often hears phrases like "collective decision" or "team decision." While collaboration is valuable, these phrases can mask reality. Often, everyone was involved, so no one is truly accountable. If the outcome is poor, each participant can reasonably say, "Well, my part was approved, and there were other factors outside my control." The result is that learning from mistakes becomes harder. In post-mortems, you find a dozen people each explaining the piece they oversaw, yet somehow the project failed and no individual decision-maker can be clearly identified as having dropped the ball. It's as if accountability has been smeared across a surface so widely that you can't get a firm grip on it. This phenomenon is well-known in social psychology as the diffusion of responsibility, where individuals in a group feel less personal responsibility because they assume others will step in. In organizational consensus theater, diffusion of responsibility is not just a quirk of group psychology but it's baked into the process. The process explicitly spreads decision rights and therefore spreads accountability thin.

Consider a scenario of a failed product launch under consensus theater. Marketing assumed Engineering was accountable for the feature quality; Engineering assumed Product Management was accountable for ensuring market fit; Product Management assumed all departments' approvals meant the plan was solid. When it flops, each function points to their checklist: "We did our part, we got our sign-offs." The real accountability -

who integrates all those parts into a successful whole - was lost in the cracks. In contrast, if a single product owner had clear authority and accountability, that person would be on the hook to reconcile marketing, engineering, and all facets into one plan, and to adjust or sound the alarm when things weren't adding up. Consensus theater often eliminates that singular accountability in favor of "governance by committee."

This distortion is not about people shirking duty on purpose; it's about how the system is set up. People within a heavily controlled process become overseers of their narrow domain rather than owners of outcomes. Their incentive is to ensure that, for their slice of the decision, nothing goes wrong (and that they have documentation to prove due diligence). They are accountable for the inputs, but not the integrated output. If something does go wrong, the natural response is to say, "It wasn't specifically my decision; we decided it as a group." In a way, the system encourages individuals to prioritize avoiding blame over achieving success, because avoiding blame has been systematized via process, whereas achieving success requires navigating a maze of shared decision authority.

An unfortunate feedback loop can result. When accountability is blurred and outcomes suffer, the typical organizational response might be to add even more reviews and approvals (in an attempt to "make sure this doesn't happen again"). That, of course, only diffuses accountability further. It's a vicious cycle: a failure occurs, no single accountable point can be identified, so the remedy is to involve even more people next time, which makes accountability even less clear. Over years, this leads to a culture where accountability feels collective and thus ethereal. People start to perceive that promotions and penalties are rarely tied to clear outcome ownership, but rather to being perceived as a diligent participant in the machine. Success has many fathers (and mothers) in the org chart, and failure, as the saying goes, becomes an orphan.

We should stress that accountability distortion is not about bad intent or lack of leadership skill. It's an emergent property of a system overloaded with process. Even excellent managers find themselves constrained. They may want to take bold responsibility but are hemmed in by the need to obtain consensus from multiple committees. The structure effectively says: "No single neck sticks out." This might protect individuals in the short run, but it leaves the organization weaker. An organization without clear accountability is like a body without a strong skeleton: it becomes amorphous, unable to stand upright under pressure.

When everyone is involved in a decision, it often means no single person feels fully answerable. We saw this clearly in our discussion of consensus theater's root causes: if "everyone is responsible" then, psychologically, no one truly is. People may even avoid taking initiative to fix issues, because doing so might mean stepping outside their swim lane in the process, which the culture has informally discouraged. Over time, talent within the organization can become demoralized. High-performing employees who thrive on accountability and decisiveness might leave, frustrated that they are not allowed to exercise those strengths. Those who remain adapt by becoming process experts and consensus-builders, valuable skills to be sure, but the organization loses the sharp edge of clear ownership.

The distortion of accountability also hampers organizational learning. When a project or decision turns out poorly, how does an organization learn and improve? Ideally, by identifying what went wrong and who might have misjudged, not to punish but to adjust course next time. In a diffused accountability environment, it's hard to pinpoint what the decision even was, let alone who made it. Lessons learned become vague: "we need better communication" or "we should involve X department earlier next time." These are hard to act on because they are so generalized. By contrast, in a scenario with clear accountability, the post-mortem might conclude something concrete like, "The launch failed because our assumption about customer need was wrong, and we know exactly how that decision was made and by whom, so that person/team can adjust their approach next time." Consensus theater's fog makes everything a collective mishap, which often means no specific corrective action is taken, aside from adding more layers (which, as we noted, can make things worse).

Consensus theater takes the sharp lines of accountability and blurs them into a mosaic of participation. Outcomes become the product of a hundred inputs, signatures, and nods, which sounds collaborative, but in practice it means no one's singular imprint or commitment is on the line. The organization as a whole becomes less adaptable because it's unclear who has the authority to change direction when needed (everyone has a little authority, but no one has enough). And when adaptation does happen, it's slow and arduous, requiring mustering the same diffuse coalition that made the decisions in the first place. This weakened accountability is arguably the gravest risk of the illusion of control we've been examining. It's the illusion that the system is accountable (because so many people reviewed it) when in fact accountability has quietly evaporated.

By stepping back to synthesize these points, we can see the full picture of how a well-meaning organization can trap itself in an illusion of control. What began as rational responses to human limits and coordination challenges, adding process, oversight, and shared decision-making, evolves into a state where the forms of control are everywhere but actual control is lacking. Consensus theater emerges as a rational but ultimately counterproductive response to rising coordination costs and fragmented risk. It creates the appearance of diligence and safety, yet often reduces the organization's real influence over outcomes. Importantly, this is not a moral indictment of the individuals involved. We have described system dynamics, not personal failures. The managers drowning in dashboards, the committee members rubber-stamping decisions, the executives asking for one more round of approvals, they are operating logically within the environment that's been built. The structure itself is the culprit that puts good people in a position where they feel they must perform these rituals.

Recognizing the patterns of consensus theater and the illusion of control is the first step toward reclaiming true effectiveness. It allows us to ask: what would it look like to design our coordination and control mechanisms so that visibility informs action rather than substituting for it; so that process aids judgment rather than obscuring it; and so that accountability is sharpened rather than diffused? These questions point to a shift in how we might run organizations, a shift from heavy process to clearer principles, from endless approvals to defined decision rights, and from comforting rituals to authentic responsibility. In fact, many organizations are already trying to solve these issues by leveraging technology - digitizing workflows, automating approvals, and using collaboration software - in hopes that this will streamline coordination and restore clarity. Digitization can certainly help up to a point, but it does not automatically break the illusion if it simply encodes the same old habits in digital form. In the next chapter, we will examine why digitizing the decision process, moving our reviews, reports, and approvals into software, often feels like progress yet fails to resolve the underlying problems of consensus theater. Understanding that will set the stage for how organizations can genuinely regain control in an age beyond process-as-usual.

WHAT THIS CHAPTER ESTABLISHES

This chapter establishes that when coordination becomes expensive and uncertainty rises, many organizations respond by expanding control in ways that feel responsible but often reduce real effectiveness. More reviews, more approvals, more reporting, and more governance forums are rarely introduced out of vanity. They are introduced to restore confidence. The problem is that confidence and control are not the same thing.

The chapter also establishes two tightly linked illusions. The first is that visibility is control. Seeing more of the work, measuring more of the work, and discussing more of the work can create the impression that outcomes are being actively managed, even when the organization's actual ability to influence those outcomes is weakening. The second illusion is that activity is progress. When projects are surrounded by constant updates, dashboards, and recurring meetings, the organization can appear highly diligent while remaining stalled in execution.

From there, the chapter explains how decision proliferation becomes a structural consequence of fragmented risk. Distributing expertise and accountability across functions is rational, but it can multiply decision points to the extent that decisions become slower, diluted, and difficult to own. The organization ends up with too many decision-makers and too few decision owners, which creates a vacuum at the center of responsibility.

The chapter then establishes consensus theater as the predictable endpoint of these dynamics. When decisions are hard to make cleanly and accountability is diffuse, organizations often substitute rituals for ownership. Meetings, approvals, and documentation become a form of insurance and a form of cover. They reduce anxiety, spread perceived risk, and create the appearance of diligence. Yet the safety is often illusory. It comes at the cost of decision latency, weakened learning, and distorted accountability.

This chapter establishes the key consequence: when everyone is involved, accountability becomes harder to locate and harder to enforce. Outcomes can be the product of many small approvals without any single person owning the integrated result. That weakens the organization's ability to adapt, because adaptation requires clear decisions and clear responsibility.

This sets up the next step in the argument. When organizations recognize the burden of consensus theater, many attempt to fix it by digitizing the theater of moving approvals, reporting, and governance into software. The next chapter explains why digitization improves throughput and visibility but often preserves the underlying decision dynamics, making the theater more efficient without making it less theatrical.

6

WHY DIGITIZATION WASN'T THE FIX

Digitization was supposed to be the cure for cumbersome workflows and sluggish processes. As organizations moved from paper and pen to software and systems, many assumed their problems would melt away in a tide of efficiency. In practice, however, the digital revolution largely preserved the shape of the old model it was meant to reform. Companies translated existing methods into digital formats without challenging their underlying assumptions. Work moved to screens and databases, and the pace of each task accelerated, but the fundamental blueprint of how decisions are made and how work flows remained strikingly familiar. This chapter explores why simply going digital didn't fix longstanding organizational issues, how new technology often sped up the same old processes instead of transforming them, how it gave us unprecedented visibility without improving judgment, and how automation struggles in a world full of exceptions. The goal is not to lay blame, but to understand why well-intentioned digitization efforts fell short of true change. By respecting the rationale behind established practices, we can see clearly how digital tools enhanced those practices without altering their core, and why better tools alone can't resolve structural mismatches in work and decision-making.

DIGITIZATION WITHOUT TRANSFORMATION

When organizations first embraced digitization, they did so cautiously and pragmatically. After all, the existing processes had evolved over years or decades, they were familiar, even if imperfect. So, the safest path was to replicate those processes in digital form. Early computerization projects were often described as "paving the cow path," meaning they took a winding, inefficient road and simply laid pavement over it. A publicly available U.S. government IT review noted that agencies had "computerized existing workflows, essentially, by paving the cow path", layering technology onto processes that already wandered through bureaucratic twists and turns. The result was a modern-looking system built on the bones of the old procedures. The user interface changed, screens instead of paper, clicks instead of ink signatures, but the number of steps, the decision points, and the approvals required stayed largely the same. It was digitization by direct translation rather than true transformation.

This copy-paste approach to digital "transformation" became commonplace. In reality, many organizations equated going digital with doing the same work electronically. As one commentator publicly observed, companies "just do the same they did before, but digitally," simply replicating existing processes in a digital environment. A paper form that once required three departmental stamps might become an online form requiring three sequential email approvals. The medium changed, yet the workflow remained intact. For example, a business might replace physical files with a document management system, a welcome efficiency, but if the business rules and hierarchy around those documents remain unaltered, then little has fundamentally improved. "Yes, they've gone digital, but they've merely translated their processes, not transformed them," as Sacha Storz has publicly observed, "They keep their structures, culture, processes, roles... they just use 'digital' as a new language... for saying the same things." In other words, the digital makeover made things look new on the surface while the organization's skeleton, its division of responsibilities, its approval chains, and its criteria for decisions has stayed old and familiar.

It's important to understand why this happened. Established processes represent a kind of organizational memory and stability. Completely redesigning how work gets done can feel risky and revolutionary, whereas digitizing what you already have promises quick efficiency gains without rocking the boat. Early digitization efforts did deliver clear benefits:

records became easier to search, calculations became instant, communication sped up. For many leaders, these improvements validated the strategy of changing the tools rather than the process. The existing structures had reasons behind them, controls to reduce error, approvals to manage risk, silos to develop expertise. Digitization, in their view, was a way to honor those rationales while removing obvious frictions (like lost paperwork or slow information flow). In essence, they respected the existing model and sought to make it faster and more efficient with technology. The trouble was that efficiency alone doesn't guarantee effectiveness. By largely preserving the structure of the old model, digitization often streamlined the status quo instead of fixing its flaws.

ACCELERATING WORKFLOWS, PRESERVING ASSUMPTIONS

If the first impact of digitization was to map old processes onto new platforms, the second was to speed up those processes dramatically. Digital tools excel at acceleration. A task that once took days as a folder crept through intra-office mail could be done in minutes via a routing software. Calculations that consumed hours of clerical work could update live in a spreadsheet. Reporting that used to rely on quarterly printouts turned into real-time dashboards. Every step of a workflow seemed to move faster and with less labor, which was undoubtedly a win. However, these faster workflows still rested on the same assumptions and decision logic as before. The technology did what it was told, but what it was told to do came from yesterday's thinking.

A classic insight from the early days of business computing still holds true: "Automation applied to an inefficient operation will magnify the inefficiency." If a process contained unnecessary steps or poorly defined rules, automating it didn't remove those burdens, it often enshrined them. The software would dutifully execute every legacy rule and approval requirement, sometimes at lightning speed and massive scale, which could actually make problems more visible. Bill Gates has been publicly quoted as saying that technology is a magnifier of intent and design. A flawed process, when automated, can generate flawed outputs faster and in greater volume. For instance, consider a customer service workflow built on the assumption that any refund over a certain dollar amount must get managerial sign-off. If that threshold was set decades ago and never revisited, a modern system will continue to enforce it without question, only now it might flag hundreds of refund requests per hour for manager approval, overwhelming supervisors and delaying responses. The underlying business rule (the threshold) was never reconsidered, only embedded in code. Digital speed, in this case, simply highlights how out-of-date that rule is, as managers scramble to keep up with an automated flood of approval tasks.

The problem is not the acceleration itself, faster is better, all else equal, but the lack of re-examination of why things are done the way they are. Organizations often assumed their workflow designs were sound and just needed technological enhancement. If employees once copied data from one form to another, a software integration now moves the data automatically, but perhaps no one asked whether both forms were necessary at all. If a multi-step review had been instituted to catch errors, a digital process might route items through the same gauntlet, even if error rates have plummeted or the nature of work has changed. Digitization rarely questioned these inherited practices; it mainly turbocharged them. As one business leader quipped, "If you digitize bad processes what you get is bad digital processes." The speed and precision of computers give a false sense that things must be working well, after all, tasks are being completed faster and with fewer obvious mistakes. But speed can mask deeper issues: a wasteful step done quickly is still a wasteful step, and an unnecessary approval completed via one-click in a system is still an overhead that might not add value. In many cases, companies ended up with the same assembly line, simply running on a faster engine.

To be fair, digitization did often eliminate certain obvious inefficiencies. Nobody misses waiting for files to be pulled from storage, or tallying numbers by hand. Routine work became easier, and that's no small achievement. The error rate for well-defined, repetitive tasks fell sharply, and capacity increased. The danger lies in assuming that improving parts of the workflow (the tasks) automatically improved the whole. Organizations preserved their old decision frameworks and roles, letting technology mostly tackle surface-level friction. For example, digital expense reporting cut down the time to process receipts, but it didn't necessarily revisit policies about what required approval or why the reimbursement process was so convoluted to begin with. The system might have sped up data entry and auditing, yet an employee could still be waiting on three different managers to OK a trivial expense because the policy behind it was never redesigned. The assumptions baked into the "old way", about risk, trust, control, carried over into the digital era. Only now, any pain they caused could show up more quickly and, sometimes, more painfully, since there was less delay to buffer the impact. Digitization gave us faster workflows, but faster isn't smarter. The model's core logic remained unchallenged, and if that logic was lacking, speeding it up offered only pyrrhic gains.

DATA EVERYWHERE, DECISIONS NOWHERE

One of the undeniable gifts of digitization is visibility. Where managers once flew half-blind, now they can see detailed metrics on every process, often in real time. Digital tools have equipped organizations with dashboards, analytics, and status updates at the click of a button. It's common today for a team to track dozens of performance indicators continuously, or for a customer's entire history to be available instantly to a service agent. The transparency and quantity of information is astounding compared to the paper age. Yet with this flood of data comes a new challenge: visibility increased far faster than our ability to make sense of it, or to decide based on it. Many organizations discovered that more data did not automatically translate into better judgment or clearer decision-making. In fact, it often had the opposite effect, a phenomenon sometimes called analysis paralysis.

With every process instrumented and every employee's output measured, the sheer volume of information can lead to "data blindness and confusion, rather than clarity," resulting in analysis paralysis. Workers at all levels report being inundated by reports, emails, and updates. Public reporting has highlighted that information overload has become a daily reality: nearly two-thirds of organization staff says they are often overwhelmed by the amount of data and emails they receive. What was intended as increased transparency, say, CC'ing more people on an email thread or providing enterprise-wide access to a reporting dashboard, can backfire when every day brings a torrent of new figures to interpret and messages to answer. Instead of empowering quicker decisions, the digital deluge can slow them down, as people sift through mountains of data looking for meaning. For example, a sales team given 50 metrics to watch may struggle more to pinpoint which levers truly drive performance, compared to when they had five key metrics in the past. The granularity of modern data can obscure the big picture. We can measure everything, but what deserves attention? In many cases, organizations didn't have established norms or training for turning all this new data into insight. The result: visibility for its own sake, without a corresponding increase in clarity of action.

Moreover, human judgment doesn't automatically accelerate just because information flows faster. Deciding well often requires reflection, experience, and discerning signals from noise, all things that don't compress neatly on a timeline. Digital systems, by surfacing issues immediately, can create an implicit pressure to respond immediately as well. Managers see a dashboard flashing a warning or an anomaly and feel they must react, sometimes without enough context or analysis. The organization gains a high-frequency awareness of its vital signs, but if its decision-making processes (the who, how, and when of making a call) haven't matured, it can end up in a state of perpetual, anxious monitoring. More alarms and indicators go off than the team can reasonably address, or than the old decision cadence could handle. Paradoxically, some companies have found that despite being awash in data, they struggle to make decisions with confidence. There is always another report to read, another analytic cut of the data that could be examined, so decision cycles drag on, or choices become timid, aiming to appease every metric.

Increased visibility also means everyone sees more of each other's work, which can blur accountability if not managed well. In the past, a problem might surface only when it became big enough to escalate to the next level; now issues are apparent to many observers in real time. This can be positive, enabling early intervention, but it can also lead to second-guessing and micromanagement. For instance, if an executive can now drill down into a frontline worker's daily productivity stats, they might be tempted to intervene or question minor fluctuations that would previously have been handled within the team. The clarity of who gets to decide what can suffer when so much information is visible to all. Middle managers in some organizations lament that senior leaders, armed with live data, sometimes bypass the chain of command, making calls on matters formerly delegated. The tools provided visibility, but not guidance on roles or decision rights. Suddenly many eyes are on the performance readouts, and if the company hasn't clarified "who ultimately decides and on what basis," the result can be confusion or conflict.

In short, digitization brightly lit up our enterprises. We can see everything, but that doesn't mean we automatically know what to do with what we see. Without evolved decision processes and clarity of judgment, more data can simply mean more to argue over or more reasons to hesitate. The promise was that perfect information would yield perfect decisions, and indeed, having data is far better than operating on guesswork. But data is only the raw material for decisions. Humans still have to interpret and prioritize it. Many organizations discovered that while they gained nearly X-ray vision into their operations, they did not gain an equivalent boost in decision-making clarity or speed. The visibility arrived faster than the capacity to exercise sound judgment on that visible terrain. The end result has been a kind of high-definition opacity, everything is illuminated, yet the path forward can be just as murky as before, unless the organization learns how to turn information into insight and action.

MORE PARTICIPATION, LESS RESOLUTION

If there is one area where digitization truly changed the game, it's collaboration and communication. Tools for email, messaging, video conferencing, document sharing, and workflow management have made it easier than ever to involve people in a process. In theory, this should lead to better decisions and more inclusive problem-solving: all the right experts and stakeholders can weigh in quickly regardless of location. In practice, however, many companies found that digital process tooling scaled up the "consensus theater" without necessarily resolving its core dramas. By "consensus theater," we mean the often elaborate performance surrounding decision-making in organizations, the meetings, the presentations, the rounds of input and review that give an appearance of thoroughness. Digital tools extended this theater by making it simple to add more actors, more scripts, and more showings. Unfortunately, more voices and interactions did not always equate to clearer or faster decisions; in fact, it often prolonged the play.

Consider the modern meeting. Scheduling a meeting used to require phone calls and juggling calendars; now an online calendar can coordinate dozens of people with a few clicks, and video conferencing means no one even has to leave their desk. The barrier to calling a meeting has dropped dramatically. Consequently, employees are spending an astonishing portion of their time in meetings and on email. One study in the Harvard Business Review reported that these collaborative activities "account for as much as 80% of employees' time in many organizations, leaving little time to complete essential tasks.". This is digital tooling at work: calendars, email, and chat make it trivial to loop more people into discussions, so that's exactly what happens, people get looped in by default rather than by true necessity. The result is "collaboration overload," in which time and energy for actual productive work diminish. The decision process becomes more populated and busy, but not more effective. In fact, evidence suggests that adding more cooks in the kitchen often yields worse outcomes. Publicly reported research has found that collective decision-making in many cases did not improve accuracy and took longer than decisions made by individuals. In other words, the digital ability to involve everyone can dilute responsibility and slow down resolution. What was meant to be an inclusive brainstorm or approval round can turn into a protracted consensus-seeking exercise where, ultimately, someone still has to make the call, only now it's after delays and compromises that might not have been needed.

Digital collaboration tools can also create a false sense of participation and consent. It's easier now to solicit input (send a quick poll, invite comments on a shared document, cc a distribution list for "feedback"). But if that input isn't genuinely incorporated into decisions, the process becomes a charade, a bit of theater to give the impression of wide involvement. People are quick to sense this. "Sometimes, people collaborate because they feel like it's expected of them, but the decision is already made," notes one leadership expert. In those cases, "collaboration becomes 'window dressing', which undermines trust." Digital tools unfortunately can amplify this issue: it's easy to ask for comments via an online platform with no intention of truly considering them, easier than convening an in-person forum and then ignoring attendees. So teams end up spending time in what they later recognize as performative exercises, eroding their faith in the process. Engagement without impact quickly turns into frustration. The very tools that were supposed to democratize decision-making can render it hollow if not accompanied by a culture of real empowerment or clear decision rights.

Additionally, the proliferation of communication channels (email, Slack, Teams, etc.) means decisions often play out in fragmented, stop-start conversations. A decision that might once have been made in one focused meeting now stretches over a week of chat messages, email threads, and side discussions, as digital communication allows continual, asynchronous debate. That can be a good thing when thoughtfully managed, it gives people time to reflect and contribute at their own pace. But often it just elongates the cycle, with issues revisited multiple times in different forums, especially if no single authority is steering it to conclusion. Digital process tools, by enabling constant discussion, sometimes create the expectation of constant consensus. People hesitate to make a definitive decision when they know a dozen colleagues could chime in at any moment with a differing view or an overlooked detail unearthed from some database. The consensus theater just keeps running through additional scenes.

Interestingly, this explosion of collaboration can sometimes dilute expertise and accountability. When a tough decision arises, instead of trusting the one or two subject matter experts, the default in a highly networked organization might be to convene a broad working group. As Professor Anya Johnson points out in published articles, over-collaboration can mean "decisions get made by a group of less-qualified people" while the true experts' voices carry no special weight, yielding "watered-down, democratic decisions that aren't necessarily the best outcome." In a digital setting, this is exacerbated by the ease of pulling in people from anywhere in the company, sometimes people who have only tangential knowledge but feel obliged to contribute since they were invited. The underlying assumption of the old model, that consensus is safe and conflict is risky, remains unchallenged; the tools simply make pursuing consensus more convenient. Rather than resolving the ambiguities of decision-making, digital tools often scaled up the audience and the noise. The stage got bigger, the cast larger, and the scripts more complicated, but the plot, how do we decide and who has the final say, stayed as muddled as ever if the organization hadn't clarified those fundamentals beforehand.

None of this is to say that digital collaboration tools are bad. They are powerful and necessary, especially as work became more distributed and complex. The issue is that the tools were layered on top of decision-making models that were never redesigned for them. The old model might have been a top-down hierarchy or a matrix or a consensus culture; whatever it was, digitization amplified its tendencies. If you had a top-down culture, now leaders can micromanage in real time with digital oversight. If you had a consensus culture, now you can involve everyone in everything and double the consensus-building time. The net effect is that the theater of decision-making, the appearances and motions of reaching decisions, expanded, while the clarity and efficacy of those decisions did not necessarily improve in step. Many organizations found themselves with more meetings, more cc'd emails, more shared documents, and still the same indecision or muddled outcomes as before, just arriving after greater effort. Digitization fixed the logistics of getting people "in the room," but not the principles of how the organization should balance participation with decisiveness.

AUTOMATION IN A WORLD OF EXCEPTIONS

One of the most touted benefits of digitization has been process automation. If computers can follow rules, then let them handle the routine, repetitive tasks, freeing humans for higher-value work. Over the past decade, technologies like robotic process automation (RPA) promised to automate anything that could be explicitly defined in steps. Many organizations eagerly deployed scripts and bots to take over mundane data entry, form processing, or transaction handling. The expectation was an immediate boost in efficiency and accuracy. And indeed, in highly structured, predictable processes, automation delivered. But a sobering lesson emerged alongside the successes: automation struggles mightily when confronted with the messy reality of exception-heavy work. Wherever a process did not fit neatly into an "if-then" flowchart, which is to say, most real-world operations, automated solutions either broke down or became enormously complex trying to anticipate every variant. This led to a proliferation of workarounds and a heavy maintenance burden to keep automations running.

The truth is that most real-world processes involve edge cases, judgment calls, or context-dependent decisions. RPA tools generally couldn't adapt to these nuances. They excelled at handling the happy path, the standard case that might constitute, say, 80% of transactions. But the other 20% would feature slight deviations: a form with an unusual field entry, a customer request that didn't match any predefined category, an invoice with an exception. In a traditional process, an experienced employee would notice the oddity and handle it appropriately, perhaps consulting a colleague or exercising personal judgment. A bot, however, has no intuition; if it encounters something it wasn't explicitly programmed to handle, it either fails or flags the item for human intervention. As one analysis noted wryly, "One exception, and the whole process breaks." Organizations found themselves building decision trees of dizzying complexity to try to cover every conceivable scenario so the bot wouldn't crash. Each time a bot failed, it required a person to step in, figure out what went wrong, and often update the bot's script to account for the new case. The maintenance of these automations became a job in itself, sometimes nearly as burdensome as the original tasks they aimed to eliminate.

A vivid example comes from industries like insurance and finance, where processes are nominally rule-based but in practice have many special cases. An RPA bot might be introduced to automatically approve straightforward insurance claims below a certain value. It works fine until it encounters a claim with an outdated form version or a missing piece of data, something a human clerk would have caught and resolved by simply contacting the agent or looking up a policy. The bot, lacking that flexibility, flags it as an error. Now a human has to handle it anyway, and additionally, someone might need to tweak the bot to recognize this scenario next time. Multiply this by hundreds of variations, and soon the support staff for the automation can seem as large as the original team of clerks. In effect, the organization has traded one kind of work for another: instead of processing claims, staff are now processing exceptions and maintaining bots. As a result, the anticipated gains shrink.

Moreover, the maintenance burden of automation in a dynamic environment is significant. Software systems change, data formats evolve, business rules get updated, and the automated scripts must keep up with all of it. One report observed that companies implementing RPA often found themselves "locked into continuous bot maintenance cycles that consumed resources and undermined the promised ROI", to the point that "the cost of maintaining the automations often rivaled the cost of the manual work they were intended to eliminate." This can become a frustrating cycle: each tweak to the automation might solve one problem but reveal another, or it might break something else unexpectedly. The more exceptions the automation tries to handle, the more intricate and fragile it becomes. Ultimately, some organizations quietly scaled back their automation ambitions, recognizing that certain jobs were too exception-laden or too reliant on human judgment to automate cost-effectively with current tools. They learned that automation is not autonomous improvement, it simply follows the recipe it's given. And if that recipe has to branch twenty different ways to account for every special case, then the simplicity and elegance that make automation attractive in the first place are lost.

There's also a human factor: when an automated system fails, employees must drop their normal work to fix things, often under time pressure. A bot might process hundreds of transactions overnight, and if it crashes on one, that item might sit untouched until a person arrives to untangle it. This creates a new kind of firefighting that didn't exist before, a fallout of trying to automate the imperfect. In some cases, staff started to build manual workarounds outside the system to handle cases the automation couldn't, creating parallel processes. For instance, if an order processing bot couldn't handle a certain product code, the team might develop a side procedure: "for product X, don't use the system, handle it via email manually." These workarounds are practical short-term fixes, but they add complexity and opportunities for error, and they must be maintained alongside the official process. Over time, the organization can end up with a tangle of automated flows, exception queues, and off-system steps, arguably more complex than the original manual process. It's a far cry from the clean, utopian vision of automation that was sold.

None of this is to say that automation has no value; on the contrary, where it fits, it excels, and repetitive drudgery has been reduced for many workers thanks to it. The key lesson learned is that automation alone doesn't resolve the deeper complexity of a process. If anything, it exposes it. Every exception is a sign that a process might not be fully understood or that the business environment can't be neatly systematized. In a manual process, experienced workers handle those ambiguities fluidly; in an automated one, each ambiguity is an error to be fixed. The takeaway is humbling: our organizations often perform a lot of unofficial adaptation and interpretation in the shadows of formal procedures. When we try to hand over work to machines, we suddenly see all the little discrepancies and judgments that humans were bridging all along. Automation hasn't yet been able to replicate that kind of flexible thinking (at least not with traditional rule-based systems). So, where work is exception-heavy, which is to say where work is truly human, automation remains an uneasy fit, leading to complexity and maintenance efforts that management must not underestimate.

BETTER TOOLS, SAME MISALIGNMENTS

Why did digitization, for all its tangible improvements, fail to fix the fundamental issues we've been discussing? The overarching answer is becoming clear: a tool, no matter how powerful, cannot by itself cure a flawed design or a misaligned structure. Digital tools are enablers; they execute the logic and processes they are given. If they are given a well-designed model, they will enhance its performance. But if they are given a problematic model, they will amplify its problems. Technology is, in a sense, agnostic, it will "faithfully and beautifully execute the underlying brokenness" if that's what exists. People who automated a bad process, as Anton Chuvakin has succinctly put it in his writing, "ended up with a bad, automated process rather than an improvement". The old structure won, and the new technology simply entrenched it. We see this in every example: digitization preserved the old workflow logic, sped it up, made it more transparent, but did not challenge its premises. The process was gravity; technology, just a rocket trying to escape that pull. And usually, gravity won, the moment the initial thrust of the digital project faded (say, once the new system was delivered and the project team disbanded), the organization's habits and structures exerted their force, bending the tool to serve the same patterns as before.

It's telling how often companies have bought cutting-edge platforms only to recreate their familiar workflows within them. A sophisticated project management suite might be acquired with dreams of agile new ways of working, yet within months the team configures it to mirror the old phase-gate approval process they've always used. An enterprise social network is rolled out to spark free-form knowledge sharing, but employees stick to official channels and hierarchies in how they communicate on it, fearing repercussions of speaking too openly. "The most common outcome of a new technology purchase is not process transformation, but forcing the new tool to mimic the old, broken workflow," as one observer noted from hard experience. This happens partly because people are comfortable with what they know and partly because organizational incentives often reward sticking with the tried-and-true. If a company's culture and structure haven't embraced new assumptions, like trusting employees with more autonomy, or empowering cross-functional teams, then even the best tools won't create those outcomes. Instead, the tools get bent to fit the existing mold.

Another reason better tooling doesn't solve structural mismatches is that problems in organizations are often social, not technical. A broken process can be a symptom of fragmented accountability, competing objectives between departments, or fear of making mistakes in a blame-happy culture. These are people problems, issues of culture, trust, and design of roles. No software can magically fix a lack of clarity in who is responsible for what, or make siloed departments share information if they're not incented to do so. "Technology does not solve human problems. A broken process is often a symptom of a dysfunctional culture," Chuvakin reminds us bluntly. You might install a powerful workflow system that can route tasks any which way, but if department A still doesn't trust department B, they'll configure that system to include every sign-off and checkpoint imaginable to protect their interests. The result? The same cumbersome process, now running on expensive software. The structural mismatch, say, misaligned goals or lack of trust, remains untouched by the tool. In fact, when the fancy new tool fails to deliver miracles, people often blame the tool itself, not realizing the tool is merely exposing or automating the contradictions they've lived with all along.

Tools tend to work at the surface level of execution, whereas fixing structural issues requires changes at the level of design and policy. For instance, if customer service is slow because reps need approvals from three different managers for an exception, implementing a faster communication system between reps and managers might shave some time, but it doesn't question whether three approvals are really needed. That question can only be answered by revisiting risk policies or empowerment guidelines, essentially, management decisions about structure and authority. Technology doesn't inherently prompt those conversations; human leadership does. Without that, tech will simply accelerate the three-approval process (perhaps making it two days instead of four), but it won't turn it into a one-approval or zero-approval process unless someone consciously redesigns the work. We saw time and again in digitization that structure trumps tool. A rigid functional hierarchy will not suddenly become nimble just because you introduce collaboration software; it will likely use that software in a top-down manner. A convoluted process will not become simple just because it's coded into an app; it will simply be a convoluted app.

Digitization didn't fix the big issues because it was, by and large, aimed at the symptoms rather than the root causes. Organizations understandably went after low-hanging fruit, speed up this task, automate that calculation, give managers a dashboard, and those are good improvements. But they stopped short of rethinking the machine itself. Better tools were bolted onto the same chassis. When those tools failed to deliver a breakthrough, it became evident that the underlying model, the who, what, when, and why of the work, was the real determinant of performance. You can tighten a few screws on a contraption and make it run a bit smoother, but if the contraption is fundamentally misaligned for its purpose, the improvement will plateau quickly. The enduring lesson is that technology, no matter how advanced, is an amplifier, not a substitute for structural wisdom. It can illuminate problems, accelerate workflows, and connect people, but it cannot, by itself, decide which processes are worth doing or how decisions should be made. Those are questions of design and judgment that remain the province of human leadership.

As we close this chapter, we have a clearer picture of why digitization, for all the genuine gains it brought, did not inherently solve the deeper challenges in organizational life. It preserved the old model's structure, accelerated its steps without challenging its rationale, flooded us with data that outpaced our decision processes, expanded the stage for decision-making without rewriting the script, and hit a wall when trying to automate work that defies rigid rules. In the end, better tools didn't resolve structural mismatches because they were never meant to; they are tools, after all, wielded within whatever framework we give them. If that framework is outdated or ill-suited, no amount of digital polish will set things right. Understanding this is not cause for despair but for hope: it means the power to truly fix things still lies in our hands, in redesigning how we organize and decide, not just in the technologies we deploy. The next steps, however, belong to another chapter. For now, it's enough to recognize the gap between digitization's promise and its reality, so that we can address the root causes of why things remain stuck despite all our new gadgets and software. Only by doing so can we move beyond simply making the old model faster, and instead begin to chart a fundamentally better course.

WHAT THIS CHAPTER ESTABLISHES

This chapter establishes that digitization improved speed and visibility without changing the underlying organizational model. In most cases, organizations translated existing workflows into software rather than redesigning the assumptions embedded in those workflows. The result was faster movement through the same constraints, not a different way of working.

It also establishes that technology tends to amplify whatever structure it is given. When a process is inefficient or overly controlled, digitizing it does not eliminate the inefficiency or the over-control. It makes them execute more quickly and at greater scale. The organization experiences more throughput, more tracking, and more reporting, but the same bottlenecks remain because decision rights and accountability have not changed.

The chapter further establishes that visibility is not the same as clarity. Digital systems produced an abundance of data, dashboards, and real-time signals. Yet decision-making did not improve at the same pace. In many organizations, more information created more monitoring, more debate, and more hesitation, rather than better judgment and faster resolution.

It also establishes that digitization often scales consensus theater. Tools make it easier to involve more people, schedule more meetings, collect more feedback, and create more artifacts that demonstrate diligence. Participation increases, but without explicit decision ownership, resolution does not. In that environment, the appearance of governance expands while accountability becomes harder to locate.

This chapter establishes that automation in a world of constant exceptions creates its own form of complexity. Rule-based automation can handle standard flows, but exception-heavy reality forces workarounds, maintenance cycles, and parallel processes. Instead of reducing burden, automation can shift it into exception handling and system upkeep, while leaving the deeper structural issues intact.

This chapter completes the diagnosis phase of the book. It explains why incremental improvement and digitization did not restore balance once coordination and control overhead became dominant. The implication is not that digitization failed, but that it succeeded within the boundaries it was given.

If digitization preserved the old model, then meaningful change must occur at the level of control itself. The next step is to examine how organizations can move from controlling work through prescribed steps to controlling work through explicit boundaries. That is the shift from process to policy.

7

FROM PROCESS TO POLICY

Process and policy are both tools for keeping work on track, but they tackle different challenges. A process prescribes specific steps and sequences for getting something done, essentially a detailed recipe or checklist telling people how to do a task. A policy, by contrast, sets out boundaries or conditions that define what outcome must (or must not) be achieved, without dictating the exact method. In simple terms, process focuses on actions, while policy focuses on outcomes. Neither is inherently better; each exists to solve a different kind of problem.

Historically, organizations relied heavily on process because it addressed a fundamental reality: human execution is variable. As discussed earlier, two people performing the same job can get very different results, and even a skilled individual can have inconsistent output day to day. Processes emerged as a way to ensure consistency and reduce risk by minimizing discretion at the point of execution. By prescribing how work should be done step by step, a process makes outcomes more predictable. For example, consider a company's purchasing practice. Instead of trusting every manager to use their own judgment for large expenses, the company might establish a step-by-step purchasing process: submit a requisition form, obtain approval from a supervisor, then from the finance department, and so on. Each step adds a checkpoint to catch errors or oversights. The process solves the problem of human variability by forcing a standard method that everyone must follow. It trades a bit of flexibility for a lot of consistency.

A policy is designed to solve a different problem: defining acceptable outcomes and boundaries when you can trust that execution will be relatively consistent. Policies become powerful when an organization is confident that tasks can be carried out without constant supervision. Instead of micromanaging each action, leadership can specify guardrails and objectives and let employees or systems operate within them. For instance, a company might have a travel expense policy that says, "Employees can approve their own business travel expenses up to $500 without further authorization, as long as they stay within budget." This rule sets a clear boundary on spending (no more than $500) but does not spell out how to book travel or which forms to fill. The details of execution, how the employee finds a flight, what booking tool they use, are left up to them, as long as the outcome stays within the allowed limit. The policy thus maintains control (by capping the amount and ensuring budget discipline) while streamlining the process for routine cases.

In practice, organizations have always used a mix of processes and policies. A bank, for example, might have a credit policy that limits how much a branch manager can lend to a single customer (an outcome constraint), alongside a detailed process for how loan officers must evaluate loan applications (a procedural guide). The policy sets the boundaries ("loans above this amount need headquarters' sign-off"), and the process provides the method ("collect these documents, perform these credit checks"). Both approaches aim to ensure sound lending; they just do it in different ways. The process intervenes in execution by guiding each action, whereas the policy frames the decision by laying down conditions.

Each approach mitigates different risks. Process guards against human inconsistency and error in the act of doing work. Policy guards against undesired outcomes or decisions that could harm the organization. Neither approach is arbitrary; both came from real needs. Earlier in the book we noted that processes became dominant for good reason: they were a reliable answer to the unpredictability of human behavior at scale. Policies, meanwhile, tended to stay in the background, broad rules that set overall limits, while processes took care of the day-to-day details.

Now, however, the balance is starting to shift. As execution becomes more reliable and uniform, due to better systems, automation, and improved training, it is less necessary to prescribe every step. When we can trust that work will be executed correctly, controlling the boundaries (through policy) often matters more than detailing the steps. In other words, if the task performance is consistently good, the organization can afford to give people or systems more leeway in how to achieve the outcome, intervening only to define the limits and required results. This doesn't make process irrelevant, but it does change its role. In an environment of high reliability, we can rely on policy to govern what must happen, and use process more sparingly, perhaps as a recommended path rather than an ironclad rule. The more execution can be trusted, the more control can shift from the minute details toward clear definitions of outcomes and boundaries.

WHY PROCESS WAS THE ONLY VIABLE OPTION

Before considering policy as an alternative, remember why process dominated for so long. In the past, detailed process wasn't just one way to manage work, it was often the only way to get consistent results given the limits of human execution. As discussed in Chapter 1, humans bring creativity and judgment to their work, but they also bring inconsistency. Ten people performing the "same" task in ten different ways will yield uneven quality and unpredictable timing. Even a single expert can have off days or overlook details. To deliver reliable results at scale, organizations turned to well-defined processes as the great equalizer. A process hard-coded the best-known method so that each person, regardless of individual habits or mood, followed the same playbook every time.

In an era before highly reliable automation, there was no real alternative to process if you needed to minimize variation. Consider manufacturing assembly lines in the early twentieth century. Managers found that if each worker used personal judgment about how to assemble a part, error rates were high and products came out inconsistent. The solution was to formalize each task, sometimes down to the exact motions, so outcomes became uniform. The same logic applied in office work. If customer service representatives handled complaints however they saw fit, one customer might get a refund and apology while another with an identical issue got a brusque dismissal. Companies solved this by scripting the interaction: every representative had to follow the same steps. The process acted as a safety net, catching potential mistakes by removing most of the guesswork and variation from each interaction.

One might ask, why not just set broad policies back then and let people improvise? The problem is that without reliable execution, a policy alone often leads to confusion or inconsistency. If management only declared a policy like "resolve every customer complaint fairly," each employee might define "fairly" in their own way, and many would be unsure how to satisfy that vague goal. Without clear steps, some employees would inevitably make mistakes that upset customers or break rules. Others, unsure of the limits, would constantly seek managerial approval ("Do I have authority to do this?"), creating bottlenecks. In short, a policy without an accompanying process would push tough judgment calls onto individuals without giving them the structure to handle those decisions consistently. The likely result would be chaos (if everyone does something different) or paralysis (if nobody feels safe acting). Thus, when humans had to make most decisions on the fly, detailed processes were essential to reduce risk. Processes didn't eliminate all risk, but they shifted much of it onto the process designers, the people who tried to anticipate issues and bake solutions into the procedure, rather than leaving every frontline worker to figure things out afresh.

Process-centric design also fostered a culture of expertise centered on following those processes. Chapter 3 noted how professionals often took pride in mastering the procedures of their domain. A veteran payroll manager, for example, wasn't just knowledgeable about payroll rules, they knew their organization's payroll process inside and out, and that mastery kept payroll running like clockwork. Clear processes also made training new employees easier: novices could follow checklists and scripts until they gained enough experience to handle exceptions with judgment. Over decades, organizations got very good at building process manuals and continuously refining them to cover more scenarios. The collective mindset became "if it's important, we'll create a procedure for it," and that served organizations well when variability was the chief enemy of quality.

This history explains why many companies still reflexively reach for more process whenever something goes wrong. If an error or lapse occurred, the typical fix was to add another checkpoint or approval step to prevent it in the future. That approach made sense under past conditions: more eyes and steps meant more chances to catch mistakes. However, the conditions that once made process king are changing. Consistency no longer depends entirely on vigilant supervision of each human action. We now have technology and systems that perform many tasks the same way every time, and better tools to guide people in their work. As a result, the need to script every detail is not as absolute as it used to be. Process is still important, but it's no longer the only way to ensure things are done right. The door is opening for policy to play a larger role, because in more and more cases we can trust the work to be executed correctly without micromanaging every step. In other words, process was the only viable option when execution was highly variable, but as execution becomes more dependable, we can manage through clear principles and boundaries rather than through exhaustive checklists.

POLICY GOVERNS OUTCOMES, NOT STEPS

If process is about the "how," policy is about the "what." A policy governs by setting out what must be true at the end of an activity or what boundaries must not be crossed during it. It defines the outcomes or conditions that the organization cares about, rather than prescribing the exact procedure. In a policy-driven approach, the organization communicates its intent, its risk tolerance, and its non-negotiable constraints up front. Execution is then expected to align with those parameters, but the exact path can vary as needed. In essence, policy tells people (or systems) what goal to achieve or what line not to cross, and leaves the how to their discretion.

Consider a few examples of policy in contrast to process. In a finance context, a spending policy might stipulate that "any purchase over $10,000 requires approval from the Finance Director." This is an outcome-focused rule: it doesn't describe how to request approval or which form to use, it simply sets a boundary on spending authority. An employee who needs to buy something knows the limit; if the amount is $9,500 and budgeted, they can proceed directly, but at $10,500 they must route it to the Director. The policy ensures control (large expenditures get oversight) without dictating the entire purchasing process. In an IT context, a data security policy might state, "Customer data must never be stored on an unencrypted device." This policy defines a clear constraint. It doesn't say how to encrypt or where to store data; those are implementation details that IT staff or software can decide. The key is that it draws a bright line: any device holding customer data must use encryption. People can choose their tools and methods, but the outcome (data is protected in a certain way) is governed by the policy.

A well-crafted policy usually makes three things explicit: the intent (the purpose or goal behind the rule), the acceptable risk tolerance (what level of variation or risk is allowed), and the specific constraints or requirements that must be respected. For instance, in the spending policy example above, the intent might be to maintain financial discipline; the risk tolerance is that small expenditures can go unreviewed because the risk is low, whereas large ones cannot; and the constraint is the $10,000 approval threshold. All of that is baked into one concise rule. By focusing on intent, risk, and constraints, policies give clear guidance without micromanaging tactics.

One advantage of governing by policy is that it enables bounded autonomy. Within the bounds of a clear policy, individuals and teams have the freedom to exercise judgment and find the best way to meet the objectives. For example, imagine a delivery operations team charged with improving on-time delivery. Management could issue a thick procedural manual trying to script every aspect of how drivers should plan their routes and schedules. Alternatively, they could set a simple policy target: "At least 95% of deliveries must be on time each day," along with any non-negotiable constraints like safety rules or customer communication standards. That policy defines the outcome (a high on-time percentage) and essential boundaries (don't break safety laws, keep customers informed), but it doesn't tell drivers exactly how to sequence their stops or which streets to take. Given this clarity, each driver or local manager can adapt and innovate to hit the goal. One driver might reorganize her route based on live traffic data; another might schedule an extra pickup during a slack period, and a third might coordinate with a colleague to swap deliveries if it makes the routes more efficient. The policy encourages these creative solutions by focusing everyone on what needs to happen, not prescribing how they must do it. The result is often better performance, because people closest to the work can apply their insight, while the organization still gets the outcome it cares about.

Policies shift the question for employees from "Did I follow all the steps?" to "Did I achieve the intended result within the rules?" This can be very empowering. It respects the capability of the people doing the work and allows them to optimize their approach. Importantly, this flexibility doesn't mean anything goes, it only works within well-defined limits. The policy provides those limits. As long as you stay within them, you have autonomy; if you need to go beyond them, that's when you stop and seek explicit approval or a policy exception.

It's crucial to stress that policy-based control is not a free-for-all, and it's not a vague suggestion to "just use good judgment." Good policies are explicit and actionable. A vague directive like "Be careful with customer data" is not a usable policy, because it leaves too much room for interpretation. What does "careful" mean? One employee might think it's fine to email a file as long as it's marked confidential, while another might refuse to share data at all because they aren't sure what's allowed. By contrast, a concrete policy such as "Do not share customer data outside the company unless an authorized non-disclosure agreement is in place" gives a clear rule that anyone can follow. If a request comes to share customer information, an employee knows an NDA is a hard requirement, no guesswork needed. The strength of policy as a mechanism lies in this clarity. When policies are unambiguous, people don't have to constantly ask for guidance or approval. They know the boundaries and can act decisively within them.

Policy focuses attention on decisions and end states rather than activities. It is a more precise form of control when execution is reliable. Instead of issuing a hundred-step procedure to guarantee a result, management can issue a handful of well-chosen policies that define what result is required and what lines cannot be crossed. As long as those policies are followed, the outcomes should meet the organization's needs. And because people (or automated processes) are free to determine the best way to achieve the outcome, the organization can become more agile and innovative without sacrificing oversight. When policy is made explicit and trusted, fewer detailed steps need to be dictated. We still get what we need (and avoid what we fear), but we do so with less friction.

WHY POLICY REDUCES COORDINATION WITHOUT REMOVING CONTROL

One of the biggest benefits of shifting from process to policy is the potential to cut down the heavy coordination overhead that plagues many organizations. In a process-centric approach, coordination is everywhere: meetings to align on procedures, multi-step approvals for routine decisions, and constant check-ins to make sure everyone is following the script. Chapter 4 described how these coordination costs can sometimes exceed the cost of the work itself. This overhead grew out of necessity, if you have to manage work through rigid processes, then every exception or cross-team interaction requires people to stop and confer. That's how you keep the process on track, but it's slow and costly.

A policy-driven approach can streamline much of this. When rules are clear and boundaries are set, people don't need to convene as often to decide how to handle each small situation. Decisions that once required a chain of emails or a committee approval can be made on the spot by the person closest to the issue, as long as they stay within policy. Take the example of an expenditure: under a strict process regime, even a $5,000 purchase might need three signatures and a procurement committee review, taking a week to turn around. Under a policy that says "managers can approve budgeted purchases up to $10,000 on their own," that same purchase can be decided in a few minutes by one person. The unnecessary coordination falls away because the rule is known and authority is granted. Only purchases above $10,000 trigger the extra approval, exactly at the point where leadership decided oversight is truly needed.

The key insight is that policy relocates control rather than abolishes it. Instead of controlling each action through continuous supervision, the organization controls the parameters and then monitors outcomes. It shifts from pre-approval of every step to post-verification that the outcomes stayed within bounds. In our spending example, management isn't reviewing every single purchase request in advance; instead, they've set a policy and can review spending reports after the fact. If someone abuses their authority, say an employee tries to circumvent the $10,000 limit by splitting a $20,000 order into smaller pieces, that will show up in the reports or audits, and corrective action can be taken. The control is still there, just applied differently. It's upstream in the sense that leadership defined what is acceptable, and downstream in that results are checked. What's gone are the multiple control checkpoints in the middle of the process that used to slow everything down.

This reduction in routine oversight can make the whole organization more agile. Different teams can work in parallel without constantly asking permission from one another, because they share an understanding of the rules of engagement. If policies are well communicated, everyone knows, for example, "Marketing can approve price discounts up to 15% on their own, but anything beyond that needs Finance's sign-off." With that policy in place, Marketing and Finance don't need a meeting for every single deal, Finance only gets involved when a proposed discount breaches the 15% threshold. That means fewer meetings and emails for the vast majority of cases, without losing control over the big exceptions.

It's worth acknowledging that moving to this model requires a shift in management mindset. Some managers initially feel uneasy not giving explicit approval for every action. In a traditional process setup, a manager might derive comfort from seeing and signing each step, it's tangible control. In a policy setup, that manager has to trust that the team is following the policy and that if something is amiss, the system will alert them. This demands confidence in the clarity and enforcement of the policies. In place of watching each move, leaders focus on designing good policies, training the team on them, and monitoring the outcomes. For many, this is a welcome relief, they can spend more time on strategic issues and less on micromanagement, but it does require letting go of the immediate visibility that comes with constant oversight.

The reassuring fact is that control isn't actually reduced; it's made more efficient. When everyone understands what "good" looks like (through policy) and has the tools to measure it, coordination can happen implicitly. The team aligns by adhering to the same guiding rules rather than by continuous discussion. When something does drift out of bounds, that's when coordination kicks back in, and then it's focused on the exception, not every routine transaction. Thus, policy reduces the need for day-to-day coordination while preserving the organization's ability to steer and correct course when it truly matters. It achieves a simpler, faster way of working without relaxing the standards that leaders care about. The caveat is that this only works if the policies are well-designed and clearly understood; otherwise, people will revert to asking for guidance. That is why the next crucial step is ensuring that what needs to be defined is made explicit.

WHAT MUST BE EXPLICIT FOR POLICY TO WORK

A policy-driven system can only function if certain things are made unambiguously clear. Unlike a step-by-step process, which can tolerate some fuzzy areas by relying on a supervisor's judgment, a policy has to stand on its own as a guide. If key elements are fuzzy or left unstated, people will either hesitate to act (for fear of doing the wrong thing) or they'll interpret the rules differently, undermining consistency. To avoid that, leadership must explicitly define the intent, the risk boundaries, and the constraints of any policy.

First, the intent behind each policy should be clear. People need to know why a rule exists, so they can apply it intelligently. For example, if there's a policy that "no project should exceed its budget by more than 5% without executive review," the intent is presumably to maintain financial discipline and avoid runaway costs. If teams understand that, they'll treat the 5% threshold seriously, not as an arbitrary number, but as management's tolerance for deviation. Intent guides people's judgment in the gray areas. It turns policies from blunt instruments ("don't do X") into guided tools ("avoid X because it endangers our larger goal of Y").

Next, the organization must define its risk tolerance in concrete terms. This is often the hardest part, because it forces explicit decisions about how much uncertainty or potential error the organization is willing to accept. In a process-heavy model, risk tolerance was implicitly very low, hence all the approvals and checks at every turn. In a policy-driven model, you decide up front where to draw the line. Consider a bank setting a lending policy: it might decide that "loans up to $50,000 can be approved by a branch manager without senior review if the applicant meets criteria A, B, and C." That policy is effective only because the bank has made a clear choice about risk: it will accept the small risk of those loans being approved locally, in exchange for faster service and less burden on headquarters. If the bank were uncomfortable with that risk, it would either set a lower limit or require more conditions. The point is, the threshold ($50,000) and the criteria (A, B, C) explicitly define the boundary of acceptable risk. Without such definition, employees wouldn't know their authority and would likely escalate every borderline case out of caution, nullifying the whole purpose of the policy.

Finally, constraints and requirements must be specific. Policy works by delineating what is permitted and what is not, so the details matter. If a policy says "handle customer data securely," that's too vague to be actionable. Does "securely" mean encryption is required? Does it mean only certain people can access the data? A clear policy would remove doubt: for instance, "Customer data must be encrypted when stored, and only employees in the Security group have access to the decryption key." Now the constraint is concrete. Anyone reading the policy knows exactly what it demands. There's no room for personal interpretation on that point, it's either encrypted or it's not, either you have the clearance or you don't.

We saw in Chapter 6 that many early attempts at digitizing workflows failed to simplify work, precisely because they preserved implicit or unclear rules. If a policy or rule was fuzzy, automating it didn't fix the fuzziness, it just embedded it in software, which then had to flag exceptions or ask humans for input at every uncertain point. The lesson is that going to a policy-centric approach often requires doing the hard work of clarification. This includes deciding on tolerances and thresholds that processes used to hide. It can be uncomfortable to pin down, for example, exactly how much money or risk a front-line employee is allowed to commit without approval, but that clarity is the price of moving faster later. If you skip this step, you don't really get the benefits of policy, people will end up seeking case-by-case guidance, and you're back to ad hoc process.

Many organizations have avoided making policies so explicit because process allowed them not to. As long as you had layers of supervision, you could afford to have policies that were more like broad principles than precise rules, any ambiguity would be sorted out by asking the boss or convening a meeting. But if you want to remove those layers of constant checking, you have to replace them with clarity. The trade-off for reducing day-to-day oversight is that the rules governing the work must be unambiguous. Everyone in the chain, from leaders to front-line staff, needs to understand exactly what the boundaries are and how much variation is okay.

The goal isn't to cover every imaginable scenario with a rule, that's impossible, but to cover the common and critical scenarios so that normal operations don't constantly stall on interpretive questions. If people are still frequently asking "Am I allowed to do this under the policy?" or "What does this rule really mean in situation X?", then the policy isn't detailed enough yet. The organization would need to refine it. By making expectations clear, you allow the machinery of the organization to run without constant human intervention, which was the promise that led us to consider policy over process in the first place.

WHAT DOES NOT CHANGE WHEN MOVING TO POLICY

With all this talk of changing how we control work, it's important to emphasize what stays the same. The move from process to policy alters the mechanics of control, but it does not alter the fundamental responsibilities or stakes. Accountability, for instance, remains as central as ever. If anything, making policies explicit can sharpen accountability. If something goes wrong, we still ask who was responsible for the outcome. Whether we operate by process or by policy, ultimately a person (or team) is answerable. The shift to policy does not let anyone off the hook, humans remain accountable for setting good rules and for acting within them. Likewise, oversight is still needed. The format changes, instead of supervising every step, oversight might mean monitoring outcome metrics, reviewing logs or audit reports, or conducting periodic checks that policies are being followed, but the purpose is the same: to ensure the organization's standards are being met. Leaders will still pay attention and require transparency; they'll just focus more on whether results align with policies rather than watching each task in real time.

Regulatory obligations and other external requirements also remain fully in force. If your company is in a sector with strict regulations or safety standards, those must be honored whether you use process or policy to manage work. In fact, many regulations mandate having explicit policies for certain matters (for example, risk management or data protection policies). Shifting more of your internal governance to policy is compatible with these mandates, but it doesn't remove them. You still need to demonstrate to auditors and regulators that you have control over your operations. The difference is that you might show them a set of policies and automated enforcement logs instead of thick procedure binders. But if a law says every transaction above a certain size requires sign-off by a licensed auditor, you must have a policy (and system) that ensures that happens, and records to prove it. Similarly, the organization's ethical standards and core values remain non-negotiable. Policy is a tool to uphold them, not a substitute for them. If anything, policies will explicitly encode some ethical positions ("we do not sell customer data," "we do not engage in bribery"), but the mere existence of a written policy doesn't guarantee ethical behavior. People at all levels still have to live those values. Leaders must create policies that reflect the organization's principles, and everyone must be prepared to exercise judgment in situations where a rigid rule might conflict with doing the right thing. In short, moving to a policy model doesn't give anyone a moral holiday, fairness, honesty, and other values still need to be actively safeguarded by humans.

The need for human judgment and expertise remains, albeit in a somewhat shifted capacity. We still rely on experienced professionals to design the system and to handle the situations that rules can't anticipate. An automated or policy-driven process may execute routine tasks, but humans decide what those policies should be in the first place. The deep knowledge of veterans is crucial in setting sensible rules and adjusting them when things change. And when an unusual situation arises that falls outside or between policies, it takes human judgment to resolve it wisely. Chapter 3 highlighted how experts cultivated mastery in their domains; that mastery is just as important now, even if it's applied differently. Instead of demonstrating expertise by navigating a complex process, an expert might demonstrate it by shaping effective policies or by deftly managing an exception that the rules didn't cover. Also, roles like risk managers, compliance officers, and quality assurance leads remain vital, their focus might shift to designing and auditing policies, but their mission of safeguarding the organization is unchanged. The policy-based model still relies on people to define goals and limits, to monitor performance, and to step in when judgment is required. The mechanisms of daily work might be different, but the human element is as essential as ever in steering the organization and bearing responsibility for its outcomes.

WHAT THIS CHAPTER ESTABLISHES

This chapter introduced a fundamental shift in how organizations exercise control: from process to policy. We saw that when execution can be made more consistent and dependable, it's no longer necessary to micromanage every step; instead, we can define the outcomes that must be achieved and the boundaries that must not be crossed. In other words, as reliability increases, control can be applied at the level of principles and limits rather than detailed instructions. This reflects a change in how to get predictable results: previously we relied on supervising the how at every turn, now we rely on specifying what must happen (or not happen) and trusting execution within those guardrails.

Crucially, this shift doesn't throw away the old logic, it builds on it. Policy is not a brand-new invention; organizations have always had rules and guidelines. What's new is that policy moves from the background to the foreground. It becomes the primary way we govern routine work, with process taking a supporting role. We did not reject the importance of control or accountability; we simply found a more streamlined way to achieve them under new conditions. When properly implemented, policy-driven control maintains the standards and safeguards that processes provided, but with greater agility and less day-to-day friction.

This chapter marks the first major departure from the traditional, human-limited operating model. It shows that we can preserve reliability and accountability even as we remove some of the heavy scaffolding of process. By shifting focus to outcomes and boundaries, we set the stage for further changes in how work is organized. If we no longer define each role strictly by the processes it carries out, then we must think about defining work in terms of decisions and responsibilities. Next, the focus shifts from Roles to Decisions, reimagining the basic unit of accountability now that the old process definitions are loosening.

8

FROM ROLES TO DECISIONS

Roles in an organization are far from arbitrary labels. They were deliberately designed as structural bundles of decisions, actions, and accountability to fit within the limits of human capacity. As discussed in Chapter 1, modern organizations arose to compensate for the fact that individual humans have finite attention, variable judgment, and limited ability to oversee complex work at scale. Assigning work by role was a direct response to these constraints. A role defines a clear scope of responsibility: it tells everyone who is expected to decide and act in a particular domain. For example, a finance manager's role encompasses decisions about budgeting and expenditures, while a safety officer's role covers decisions about risk protocols and compliance. Each role groups together related decisions and the actions that follow from them, ensuring those decisions are made by someone prepared and accountable for the outcomes.

This approach simplifies coordination dramatically. Instead of negotiating who will handle each task or decision on the fly, organizations rely on roles to provide default answers. Chapter 3 highlighted the importance of mastery in these roles: over time, individuals in defined roles develop deep expertise and sound judgment within their scope. That expertise makes their decisions more consistent and reliable. Equally important, clearly defined roles spare everyone from constantly worrying about responsibilities outside their own domain. When an issue arises, people know which role should take charge. Sales questions go to the head of

Sales; legal questions go to Legal Counsel. By bundling decisions and actions into stable roles, the organization minimizes ambiguity about ownership. Everyone can focus on their own work without continuously clarifying who is supposed to do what.

Roles have been indispensable for scaling up organizations. In a large enterprise, tens of thousands of decisions are made every day. It would be impossible for every decision to be debated collectively or assigned case-by-case without grinding work to a halt. Roles serve as a coordination mechanism that divides the vast decision load into manageable portions. They are not merely boxes on an org chart; they function as the basic architecture that holds the organization together under conditions of human limitation. Each role carries its own slice of accountability, which allows complex operations to run without constant central oversight. A well-defined role means that someone, by virtue of their position, is trusted to handle specific problems and decisions so others don't have to monitor or second-guess those matters continuously.

Crucially, roles also carry the authority needed to act on their decisions. The person in the role is empowered to make calls within their domain without seeking approval for every routine matter. This pairing of responsibility with control underlines why the role construct was so successful: it unified decision-making and execution. If a manufacturing supervisor is responsible for production quality, they are also given the authority to stop the production line if something is wrong. The decisions and the actions are bundled together under one accountable roof. This clarity of "who decides and who acts" reduced friction in day-to-day operations and enabled organizations to run efficiently despite human constraints.

The concept of roles as fixed bundles of decisions has served organizations well. It reduced chaos and provided much-needed clarity, allowing growth by delineating who handles what. However, as the nature of execution evolves, the inherent limitations of these fixed bundles are beginning to show. The conditions that once made strict role definitions essential are no longer as absolute as they used to be. When the way work gets done changes, be it through improved reliability, automation of certain tasks, or new forms of collaboration, the rationale for clustering decisions so tightly within traditional roles comes into question. The usefulness of roles as the sole organizing principle may begin to weaken under these new conditions, prompting a need to examine responsibility through a different lens.

WHY ROLES REDUCED COGNITIVE AND ORGANIZATIONAL LOAD

A system built on roles does more than clarify who is in charge; it also lightens the mental and coordination burdens on everyone involved. Clear role definitions narrow each person's decision space. When you know the boundaries of your role, you don't have to constantly wonder whether to weigh in on a given issue or if it's someone else's responsibility. Because each role has a defined scope, individuals can concentrate on a manageable set of decisions and not get distracted by everything happening across the organization. This reduction in ambiguity lowers cognitive load. People in well-defined roles can pursue their work confidently, trusting that other roles will handle their own domains. They aren't forced to renegotiate or clarify responsibilities at every turn.

Roles also streamline organizational communication and coordination. As noted in Chapter 4, coordination becomes costly when too many people must align on each decision. Roles mitigate this by acting as a compression mechanism: they limit how many individuals need to be consulted for any given matter. If a question arises about a client contract, everyone understands that it falls under the Legal department's scope and there is no need for a dozen different voices to chime in. By designating a primary owner for each category of decision, roles allow the majority of decisions to be made locally, within the team or function responsible, rather than climbing a ladder of approvals. This makes day-to-day operations more efficient. Each decision doesn't require assembling the whole village; most decisions can be resolved within the small circle defined by the relevant role.

This efficiency was both practical and necessary when human decision-making and execution were tightly intertwined. In the traditional model, the person making the decision often had to execute it or ensure it was executed correctly. With roles, that linkage remained intact and clear. The marketing manager who decides on a campaign approach will also oversee executing that campaign; the plant supervisor who decides to adjust a production schedule is positioned to implement the change on the shop floor. Bundling decisions and actions in one role meant fewer handoffs and less chance for miscommunication. It was a way to ensure accountability from decision through implementation: the same accountable person or team carried an initiative from start to finish, giving everyone else one point of contact to trust for that domain.

Defining roles in this way has been an elegant solution to the problem of organizational complexity. It prevented constant confusion, minimized overlapping efforts, and allowed each part of the organization to run semi-autonomously. However, the very strength of this arrangement, the tight coupling of decisions to individual humans and their immediate teams, can become a limitation as the capacity to make and execute decisions expands beyond what one person or group can handle. When new tools or processes allow decisions to be executed more automatically or information to flow instantly, the old constraints loosen. Under these emerging conditions, the rigid bundling of decisions into roles may start to create friction. What used to be an efficient shortcut can turn into an unnecessary checkpoint. In other words, as soon as decision-making and execution are no longer bottlenecked by individual human capacity, the rationale for funneling everything through fixed roles begins to weaken.

ROLES HIDE THE TRUE STRUCTURE OF WORK

Yet the convenience of roles comes with a side effect: it can obscure the true shape of the work. The reality is that any given role is responsible for a mix of distinct decisions and tasks. Some are big and infrequent, some small and routine. But when we package them all under one role, we stop distinguishing between them. We talk about "the project manager's job" or "the operations director's responsibilities" as if each were one monolithic thing. In truth, every role contains multitudes of decisions. For example, a project manager in a single week might decide how to adjust a timeline for a minor delay, approve a small budget reallocation, and also make a higher-stakes judgment about whether to recommend canceling a project that's veering off course. These decisions vary greatly in significance and risk, yet because they all fall to the project manager role, they tend to blend together as simply part of the job. The fine structure of work, the fact that it's composed of many separate choices, gets lost under the umbrella of the role.

As Chapter 7 explained, shifting from rigid processes to explicit policies changes how we perceive work. Instead of focusing only on step-by-step procedures, we begin to focus on the conditions and boundaries that define acceptable outcomes. That perspective reveals that decisions are the fundamental building blocks of those outcomes. A policy, after all, is an expression of intent or constraint. It guides decisions about what is or isn't allowed. When we manage purely by roles, we often hide this underlying decision logic. The role becomes a convenient abstraction, a shorthand for a bundle of responsibilities, rather than an exposure of each individual decision being made. This abstraction made management simpler when it aligned with the way work had to be done; it wasn't a natural unit of work so much as a useful fiction given the old constraints.

However, like any abstraction, roles are only useful as long as they mirror reality closely enough. When the nature of work or the operating environment changes significantly, the role abstraction can start to break down. A simplification that once streamlined understanding can become a source of misunderstanding. If certain decisions within a role could be handled differently, automated by software or governed by a clear policy, then treating the role as an indivisible unit obscures those opportunities to improve. In such moments, what used to be a helpful simplification turns into a constraint. Abstractions become problematic when they no longer align with reality. This is why it becomes necessary to examine work at the level of the decisions themselves, not just the broad roles that bundle those decisions together.

DECISIONS DIFFER IN RISK, FREQUENCY, AND CONSEQUENCE

A key insight in moving from a role-centric view to a decision-centric one is recognizing that not all decisions are created equal. In any given role, some decisions are rare but critical, they carry high risk or weighty consequences, while others are daily occurrences with minor impact. Many decisions fall somewhere in between. The problem is that a traditional role doesn't always make this distinction clear. All decisions within a role's scope tend to be processed in similar ways, under the same general rules and often by the same people, regardless of their significance. An urgent safety decision that could prevent a major accident, and a routine decision about reordering office supplies, might both formally belong to one manager's area and thus follow the same basic path of review or approval. From an organizational chart perspective, they're just part of "that manager's job."

This one-size-fits-all approach can lead to two opposing issues: overinvestment in trivial decisions and underinvestment in crucial ones. Because roles bundle everything together, low-risk, frequent decisions often end up receiving more attention than they truly need. Managers might find themselves convening meetings or filling out forms for matters that pose little real risk, simply because procedure or habit requires it. (For instance, some organizations still route every minor expense through a senior manager's sign-off, creating unnecessary delays for a $100 purchase.) This was illustrated in Chapter 5's discussion of "consensus theater" - the performative layer of reviews and approvals added in an attempt to maintain control. We saw that when uncertainty rises, the instinct is often to introduce more checkpoints and more observers for every decision, large or small. The result is that even routine matters bog down in process. Yet paradoxically, this added ceremony doesn't necessarily improve the quality of high-risk decisions. Having a crowd in the room or a thick packet of approval documents may create an appearance of diligence, but a rare, high-stakes decision (say, entering a new market or handling a major crisis) may still hinge on the judgment of the same one or two key people. In other words, bundling all decisions through the same role-based channels can slow down the easy stuff without materially improving the hard stuff.

By contrast, treating decisions as distinct units allows for a more precise calibration of oversight and effort. If we acknowledge that a daily $500 expenditure decision is fundamentally different from a once-a-year $5 million investment decision, we can design different handling processes for each. The frequent, low-consequence decisions might be handled automatically under well-defined policies or delegated with minimal oversight, saving time and energy. The rare, high-impact decisions can be elevated for careful deliberation, perhaps involving additional experts or stricter review, because they truly warrant that extra attention. In short, decision-centric thinking means fitting the process to the nature of the decision, rather than forcing all decisions to go through the same hoops. It provides finer-grained control: the organization can apply exactly the right level of scrutiny, speed, or caution depending on the decision's characteristics. This stands in contrast to a role-centric model, where the default tends to be a uniform treatment for everything a role touches.

Once decisions are differentiated by their risk, frequency, and importance, new possibilities for managing work open up. The organization is no longer chained to a single mode of decision-making attached to a job title or department. Instead, it can choose the best way to handle each category of decision. Low-risk, repetitive matters can be streamlined or even handled by automated systems within preset bounds, freeing human capacity. Unusual or high-stakes matters can be given the time and senior attention they demand, without the clutter of minor issues distracting from them. In this way, separating decisions by their characteristics is a first step toward a more adaptive and efficient model of governance. It reduces needless bureaucracy while strengthening control where it truly matters. It also naturally leads to the next consideration: if we manage decisions on their own terms, how do we redesign the organization's approach to assigning responsibility and authority for those decisions?

WHY DECISION-CENTRIC DESIGN REDUCES COORDINATION

Shifting the design focus from roles to decisions has a powerful effect on coordination: it localizes alignment exactly where it's needed, instead of requiring broad involvement for every issue. In a decision-centric model, each type of decision is explicitly assigned an owner and a method at the point of that decision. This means that when a decision arises, it's immediately clear how it will be made and who (or what system) will make it. There's no ad hoc scramble to figure out the chain of approval. For example, an organization might establish that any customer refund under a certain dollar amount is decided automatically by a policy (perhaps implemented in software according to rules set by Finance), whereas larger refunds trigger a review by a service manager, and only the very largest or unusual refunds escalate to an executive for sign-off. In this setup, the vast majority of refund decisions happen smoothly at the front line or via an automated workflow, without meetings or emails up the hierarchy. Only the few cases that truly warrant higher scrutiny bubble up to higher levels. The decision-making authority is tailored to the decision type. Small decisions are handled quickly at their source, and big decisions get the focused attention they deserve. The net effect is that everyone spends less time in blanket coordination mode, because "who decides what" has already been decided in the design.

This approach materially reduces the coordination tax on the organization. In the traditional role-bound scenario, even minor decisions can trigger communication across levels "let's run this by the director to be safe," or "we need three departments to sign off since this issue touches all of them." Chapter 4 highlighted how such constant alignment can become more expensive than the work itself, making progress move at the "speed of approval" rather than at the speed of actual execution. With decision-centric design, much of that overhead disappears. Decisions don't need to move up and down a long chain of command if their resolution has been pre-assigned to a specific role or automated process. People up the hierarchy no longer need to micromanage routine matters, and people down the hierarchy aren't left waiting for approvals that add little value. Coordination still happens, but it's now focused on the exceptions and the high-impact choices, not on every single transaction.

It's important to note that focusing on decisions rather than roles is not about removing structure or eliminating leadership. It doesn't "flatten" the organization into a free-for-all; rather, it channels formal authority more precisely. In a decision-centric design, hierarchy still exists, but it comes into play exactly where it should: in overseeing decisions that carry substantial risk or require broad judgment. Meanwhile, lower-risk decisions can be made with greater autonomy. In other words, control is aligned with actual risk. A well-crafted framework makes it clear which decisions can be made independently, which must follow standardized rules, and which need elevation to higher authority. This not only streamlines execution but also improves accountability. When each decision type has a defined owner and method, there is little ambiguity about who is responsible for the outcome. Responsibility is tied directly to the decision at hand, rather than being diffused across a broad job description.

By making decision rights explicit, organizations often discover gaps or overlaps that were hidden under the old role-based scheme. Perhaps two departments each assumed the other was handling a certain kind of decision, leaving it effectively unowned, or a decision was being made by multiple parties with no single accountable owner. A decision-centric mapping brings these issues to light. It forces the question: Do we know who owns this particular decision, every time it comes up? If the answer is unclear, that's a gap to be addressed. Under the role paradigm, such ambiguities could persist for a long time, only noticed when something slipped through the cracks. By contrast, when decisions are the primary units of design, it's hard for an important issue to hide in an organizational blind spot, every significant decision must be accounted for. This clarity strengthens accountability and ensures that the new structure isn't creating unintended voids. In sum, designing around decisions reduces unnecessary coordination by pre-allocating ownership, and it heightens accountability by making that ownership explicit and proportional to the decision's importance.

WHAT REMAINS HUMAN IN A DECISION-ORIENTED MODEL

All this talk about distributing and automating decisions might raise an important question: what is left for people to do? The answer, in short, is a great deal. Human expertise and judgment remain essential in a decision-oriented model, but they are applied in higher-leverage ways. People still set the guardrails and make the judgment calls that machines or predefined processes cannot. In fact, one of the most human-intensive parts of a decision-centric approach is designing the system itself: deciding which decisions can be safely automated or delegated and under what conditions. That requires deep understanding and foresight. Leaders and subject-matter experts must define the policies, thresholds, and criteria that will guide all those decisions. For example, a compliance team will articulate the rules that an automated approval system follows for routine transactions. In doing so, they rely on their experience to capture subtleties ("flag anything that looks like this pattern, allow that only up to a certain limit, and never approve those without human review"). These boundary-setting activities are inherently human and call upon exactly the kind of mastery that seasoned professionals possess.

Furthermore, humans remain the backstop for ambiguity. No matter how finely tuned a set of rules or algorithms is, the real world will produce situations that fall outside its parameters. When a novel or nuanced case arises, it is a person who must interpret the scenario and decide how to proceed. Consider an HR example: a self-service system might handle standard vacation requests automatically, but if an employee requests an unusually long leave for a unique personal situation, that decision will be escalated for a human manager's judgment. Likewise, an AI-driven analytics tool might flag an out-of-pattern market opportunity, but a human executive will weigh the context, consider risks and intangibles, and decide whether it makes strategic sense to pursue. In a decision-centric structure, those uncommon cases aren't left in limbo; they are explicitly routed to human decision-makers. This ensures that when judgment, intuition, or ethical deliberation is required, it comes into play. Rather than eliminating the need for human decision-making, the new model preserves it for where it matters most: in the gray areas that require interpretation and in the high-impact choices that define the business.

It's also worth emphasizing that high-stakes decisions continue to be a human responsibility by design. While we might automate a host of small operational choices, we do not hand over the keys when it comes to decisions that could fundamentally change the company's direction or risk profile. Top leadership will still deliberate on major investments, strategic shifts, and sensitive policy matters. What changes is that they will not be bogged down by also having to approve countless minor decisions leading up to those big moments. Chapter 3 underscored the value of mastery and experience in making sound judgments. That value does not diminish in a decision-oriented model; if anything, it becomes even more critical. Experienced individuals become the architects and overseers of the decision framework, as well as the arbiters of exceptions. They spend more of their time defining decision criteria, fine-tuning policies, reviewing aggregate outcomes, and coaching others in judgment rather than manually processing each decision one by one. Their insight guides the system as a whole and ensures that the automated or delegated decisions continue to align with the organization's goals and values.

In practice, people's roles evolve rather than disappear. A manager whose team's routine tasks are largely automated doesn't become irrelevant; instead, that manager focuses on ensuring the automated rules are correct, handling any escalations or exceptions, and coaching their team on the more complex situations that arise. Accountability remains firmly with the humans in charge. When decisions are made by automated means, it is because a human-approved policy allowed it and if that policy proves flawed, humans will adjust it. In this sense, individuals shift from executing every decision manually to curating and supervising an expanded decision-making process. They remain the ultimate stewards of outcomes. They define the intent, set the boundaries, monitor the results, and intervene as needed. They do not need to personally execute every step, but they are still responsible for ensuring that each step is executed correctly. A decision-centric design does not sideline human judgment; it elevates it. It reserves our finite human attention for the problems that truly need it, while routine matters proceed within the guardrails that we humans have set. The continuity of responsibility and oversight remains intact even as the mechanics of execution evolve.

WHAT THIS CHAPTER ESTABLISHES

This chapter has reframed how we think about responsibility in organizations. We began by recognizing that the familiar role structure was a rational and effective solution for a world in which humans were the limiting factor in execution. Roles bundled together decisions, actions, and accountability so that work could be coordinated and scaled despite our individual cognitive and physical constraints. That model still underpins most organizations today, and for good reason: it provided clarity and control in an era when every decision had to be made and carried out by people. However, we have also seen where that model obscures and constrains the work. By treating the role as the basic unit of organization, we often introduce hidden inefficiencies, for example, unnecessary coordination, blurred ownership of specific outcomes, and a one-size-fits-all approach to problems of very different magnitude. In a changing environment, where execution can be more distributed and assisted by reliable systems, those inefficiencies become more pronounced.

Shifting focus from roles to decisions allows us to refine organizational design to better match these new realities. Rather than assuming a one-bundle-fits-all approach to responsibility, we can unpack work into its constituent decisions and manage each on its own terms. This doesn't discard the structures of accountability and expertise that organizations have built up; instead, it makes them more precise. Policy can be applied exactly where it's needed, oversight can be targeted to the highest-risk areas, and execution can be accelerated in low-risk areas, all without breaking the chain of accountability. In essence, we retain the benefits of hierarchy and specialization, but we deploy them more thoughtfully. A hierarchy can now focus on providing intent and setting boundaries, while day-to-day decisions align with those boundaries independently. Accountability remains as crucial as ever, but it is tied to clearly defined decision domains and outcomes rather than to broad job titles alone.

Crucially, this is an evolution of our organizational model, not a rejection of it. We are preserving the wisdom of the existing system, the need for clarity, coordination, and human judgment, while relieving some of the strain that system is beginning to show. We are not abolishing roles or hierarchy; people will still have titles, teams, and domains of expertise. But those domains will be defined more explicitly by the decisions they encompass and influence, rather than by an exhaustive list of tasks. The organization as a whole becomes more dynamic and responsive, able to handle different types of decisions with processes proportional to each. Meanwhile, accountability becomes crisper: when something goes wrong, it's easier to trace the issue to a specific decision and its owner (or the policy guiding it), rather than untangling a web of overlapping responsibilities. And when things go right, success can likewise be attributed to the decision-makers and frameworks that were in place, giving credit in a more pinpointed way.

By establishing a decision-centric perspective, we have set the stage for the next fundamental shift in how responsibility is exercised. With decisions unbundled from traditional roles and many routine choices happening autonomously within clear guidelines, we must consider how leaders can effectively guide and influence this kind of system. How should accountability and control be asserted when oversight is no longer about reviewing every step? The answer lies in a different expression of leadership, one that relies less on inspecting each action and more on communicating a clear intent that shapes decisions at every level. In other words, responsibility moves upstream: defining purpose, boundaries, and criteria so that those distributed decisions are made in the right spirit. This structural shift, from supervision to intent, forms the next stage of our evolution. It preserves the core of accountability but changes its expression, allowing an organization to maintain alignment and trust even as it grants more freedom in execution.

9

FROM SUPERVISION TO INTENT

For as long as organizations have existed, those accountable for outcomes have felt a duty to oversee how work gets done. This instinct for supervision emerged naturally when human effort was the primary execution layer of operations. If a manager was responsible for results, the safest way to fulfill that responsibility was to watch over the work as it happened, to review, approve, coach, correct, and intervene as needed. Classic management theory formalized this oversight role as the "controlling" function of management: monitoring performance, comparing it against standards, and taking corrective action when necessary. In practice, a bank supervisor might double-check transactions, a factory foreman might inspect each batch of output, and a legal chief might require reviewing every contract. These are all forms of supervision. They arose not from mistrust or bureaucracy for its own sake, but from a rational need to make outcomes reliable when humans (with all their variability) carried out every task.

It is important to recognize that supervision, in this traditional model, was a necessary and honorable expression of responsibility. Outcomes could not simply be taken for granted; someone had to ensure the work was done right. Because humans are inconsistent, prone to fatigue, distraction, or differing judgments, responsible leaders developed processes to see into the work, not just check the end results. Having visibility into how results were achieved was essential to catch errors early, enforce standards, and coach less experienced staff. In fields from healthcare to manufacturing to finance, this hands-on oversight was the only way to guarantee quality and safety. Far from being a flaw or a sign of micromanagement, supervision was the mechanism by which experienced practitioners translated their accountability into action. It was how managers kept the ship on course. Indeed, many of the safeguards people take for granted (a senior engineer signing off on a design, a senior editor reviewing an article) exist because supervision has long been synonymous with diligence.

Yet this beneficial oversight comes at a cost. Supervision is labor-intensive by design: it requires human attention at the point of execution. The more steps or decisions a process involves, the more hours of supervisory review are needed to ensure each is done correctly. This makes supervision an expensive form of control, because it literally sits in the flow of work. A manager's time must be spent checking or approving tasks rather than producing output directly. As organizations grew and work multiplied, the burden of supervision grew with it. We can begin to see a tension: if scaling up means ever more people and decisions, can responsibility always be exercised by personally overseeing each one? The very rationality of supervision, its placement at the point of action, makes us wonder whether there might be a higher-leverage point from which to exert control.

WHY SUPERVISION SCALES POORLY

However essential it is, direct supervision does not scale gracefully. An organization that doubles in size cannot simply double its oversight and expect the same efficiency. Each additional employee, each new process or transaction, adds disproportionately to the supervisory load. A conscientious manager quickly finds that reviewing everyone's work can consume more time than the work itself. Many leaders have experienced this: as their team expands, their days fill with status meetings, approvals, and check-ins, leaving little time for strategy or improvement. This isn't because they are doing something wrong; it's a structural reality. Human attention and judgment are finite resources. No matter how skilled a supervisor is, there is a limit to how many complex actions they can meaningfully monitor in real time.

Organizational research bears this out. Even in well-run companies, a single manager can effectively oversee only a certain number of direct reports before the quality of supervision suffers. A typical span of control for a supervisor handling moderately complex work is on the order of eight to ten direct reports. In more routine or highly standardized environments, one manager might manage a bit more, perhaps 15 or so people, because less oversight is needed when tasks are uniform. But these numbers illustrate the point: supervision does not expand linearly with workload. To oversee 50 frontline employees, you can't have just one supervisor; you need layers of management, which introduce their own coordination costs. Each layer filters information and adds delay between the top and bottom. If a company kept trying to supervise every action as it grew, it would either need an impractical number of managers or it would slow to a crawl under its own monitoring process.

This scaling problem has long been recognized in management. Some call it the supervisor's dilemma: beyond a certain point, a supervisor becomes a bottleneck rather than a guarantor of quality. The best managers mitigate this by delegating some authority, but delegation itself traditionally requires trust that must be earned through... more supervision. In a growing operation, the volume of checks, approvals, and reviews can balloon until managing the process overtakes the actual productive work. We see this in anecdotes of executives who lament spending entire days "checking other people's work" instead of focusing on bigger goals. It's not a personal failing on their part; it's the inevitable result of scaling a model where responsibility is exercised through direct oversight. Eventually, organizations reach a crossroads. When supervision itself becomes the constraint on throughput and agility, leaders face a choice: either slow down to stay safe, or find a new way to exercise their responsibility at scale.

Historically, many organizations simply added hierarchy to cope with growth, creating middle-management roles to divide the supervisory burden. This worked up to a point, but it also created distance between the top decision-makers and the front line. More layers meant more handoffs and a higher chance that context or nuance would be lost along the way. And even with many managers, certain processes could still choke on the need for approvals and sign-offs at every step. The underlying issue is that traditional supervision relies on human control of each decision, and humans do not scale exponentially. This is why the quest for a more scalable form of control began. Over time, forward-thinking leaders began experimenting with shifting some control upstream, guiding how work should be done before it is executed, rather than inspecting it after the fact. The idea was to relieve the burden on supervision without sacrificing accountability. That idea is encapsulated in the shift from direct supervision to guiding intent.

INTENT IS A HIGHER-LEVERAGE FORM OF CONTROL

What do we mean by "intent" in this context? Guiding by intent means articulating a clear purpose, set of constraints, and success criteria for a piece of work, and then letting the people or systems executing it make decisions within that framework. In essence, it is a way of exercising responsibility upstream. Instead of standing over each task, the accountable leader defines what outcome is desired, what boundaries must be respected, and what trade-offs are acceptable, and trusts the execution layer to fill in the how. Intent is thus the explicit expression of what needs to be achieved and under what conditions, without dictating each step. It's a form of control that operates through clarity of direction rather than continuous oversight.

Notably, the concept of "Commander's Intent" is a cornerstone of military doctrine, providing clear guidance so that units can adapt their actions in the field while remaining aligned with the overarching mission. Similarly, "Intent-Based Leadership" has been popularized by L. David Marquet, whose approach emphasizes empowering teams by communicating objectives and boundaries, then allowing individuals to determine the best methods for achieving those goals.

Crucially, intent is not a vague mission statement or inspirational vision. It must be concrete enough to guide consistent decisions in the field. A useful statement of intent includes specifics: the goal to be reached, the parameters (like budget, timing, quality thresholds, or ethical constraints), and any priorities or preferences among objectives. In short, it provides the context a decision-maker needs to act independently in alignment with the leader's expectations.

Leading by intent does not remove accountability, it recenters it. The responsible leader remains on the hook for outcomes just as much as before. However, their accountability is now expressed by setting the initial conditions and expectations, rather than by micromanaging execution. If the outcomes go astray, it becomes a question of either the intent being flawed or not properly followed, rather than a failure to hover closely enough. In this way, accountability actually becomes clearer. It is tied to the outcomes and boundaries that were explicitly stated, rather than to whether someone followed a prescribed procedure to the letter. This clarity can be bracing: when a team is guided by intent, there is less room to hide behind "I did exactly what I was told, so it's not my fault the result was off." Those accountable cannot simply point to compliance with a checklist; they must examine whether the intent was sufficient and whether actions truly aligned with it. Thus, intent-driven control shifts the focus to results and principles, which often makes responsibility more transparent than a thick stack of process compliance reports ever could.

One way to appreciate the power of intent is to consider how it leverages human initiative. When people understand the why and what of what they're trying to achieve (the intent), they can apply their full creativity and judgment to the how. This is fundamentally different from being handed a list of instructions. As leadership expert David Marquet famously put it, "If you want your people to think, don't give instructions, give intent". In Marquet's experience as a U.S. Navy submarine captain, replacing orders with intent, asking his officers to state what they intended to do and why, created a ship of thinkers rather than mere doers. The psychological shift was enormous: team members took ownership of their decisions because they originated the approach within the commander's intent. Authority moved to where the information and expertise were, while alignment was maintained by the clarity of the stated goal and limits. This illustrates why guiding by intent is considered a higher-leverage form of control. A well-crafted intent can align countless downstream decisions without the leader needing to supervise each one. It is a force multiplier for leadership attention: instead of one manager watching 10 tasks, that manager sets an intent that 10 people use to guide 100 tasks. The "span" of effective influence increases, not by stretching one person thinner, but by empowering many others to act in harmony.

It's worth noting that intent as a control mechanism has precedents in various fields. Military doctrine has long used the concept of "commander's intent" to allow units in the field to adapt to changing circumstances while still pursuing the overall mission. In business, methods like Management by Objectives (MBO) similarly emphasize setting clear goals and letting employees determine the best path to reach them. In both cases, the logic is the same: if execution can be made more consistent and reliable (through training, standard tools, or modern automation), then leaders no longer need to hover over every move. They can focus on defining success and constraints, the intent, and trust the execution to follow those guidelines. It is control by guidance instead of control by inspection. And when execution becomes more scalable (for instance, when tasks can be automated or handled by well-designed systems), an intent-driven approach is not only more efficient but often more effective. It taps into the intelligence of the whole team or the precision of the system, rather than bottlenecking decisions at a supervisor.

That said, leading by intent is not a hands-off laissez-faire approach. It demands more clarity and discipline up front from leaders, not less. Crafting a good intent is hard work: one must think through objectives, articulate them plainly, anticipate potential ambiguities, and define boundaries that are neither too tight nor too loose. Ambiguity in intent will simply force a reversion to micromanagement when people inevitably interpret things differently. Thus, shifting to intent requires a high standard of communication and thought from those in charge. In exchange, it reduces the need for constant intervention later. When done well, an intent-driven model means that everyone down the line knows exactly what success looks like and what freedoms or constraints they have. They shouldn't need to ask "Is this OK?" at every turn, because the intent already answered that, or provided the criteria to answer it themselves. In sum, intent as a form of control relocates effort to the planning and context-setting phase. It is there, upstream, that the leader exerts their influence most. The downstream execution then carries that imprint even when the leader isn't personally present at each decision.

FROM CHECKING WORK TO SHAPING DECISIONS

Shifting from supervision to intent also transforms the daily stance of leaders and managers. Under the old model, responsibility meant checking work, verifying after the fact (or during it) that each action was correct. Under an intent-driven model, responsibility shifts to shaping the decision environment before the fact. Instead of catching mistakes downstream, leaders focus on preventing mistakes upstream by making it easier for people to choose correctly. This involves defining clear boundaries, establishing priorities, setting escalation triggers, and ensuring that those doing the work understand how to judge success. In practical terms, it means a leader might spend less time poring over completed reports and more time refining the guidelines and training that teams use to execute their work. The effort moves from inspection to design: designing processes, policies, and contexts such that even unsupervised decisions will tend to be the right ones.

One way to illustrate this is to consider quality control. In traditional supervision-heavy systems, quality is often maintained by inspecting finished work and sending it back for rework if it's not up to standard. This is akin to catching errors at the end. But a famous principle in quality management, articulated by W. Edwards Deming, is that "you cannot inspect quality into a product. The quality is there or it isn't by the time it's inspected". Deming's point was that quality must be built into the process (through good design, clear standards, and proper training) rather than solely relying on post-hoc inspection. The same idea applies to managerial oversight broadly: you cannot "inspect" success into a project by reviewing it at the end; you have to build success into the way the project is set up from the start. Shaping decisions is about creating the conditions in which the right actions happen naturally, rather than catching the wrong actions after they occur. When leaders define a clear intent and equip their teams with sound decision criteria, they reduce the need for later corrections. Fewer things go wrong that require intervention, because many potential missteps were preempted by the structure and guidance given.

What does shaping the decision environment look like in practice? It can take many forms, all of which serve to guide execution in lieu of constant oversight. It might mean establishing policies that encode what is acceptable or preferred in common scenarios, so workers can make routine decisions without asking for approval each time. (For example, a company might have a policy that any refund under $100 can be issued by front-line customer service reps without manager sign-off, as long as certain conditions are met. That policy shapes decisions on the ground and reduces oversight needs.) It also involves setting priorities and values so that when trade-offs arise, people know which principle to favor. A leader might emphasize, for instance, that "safety comes before speed" in a manufacturing setting, this guiding intent ensures that when in doubt, employees will choose the safer route even if a supervisor isn't there to tell them. Shaping the environment includes defining escalation triggers: clearly stating the conditions when a decision should be bumped up to higher authority. This way, managers don't have to watch everything; they trust that only genuinely unusual or high-risk cases will be brought to their attention. It also means delineating success criteria in advance, how will we know a decision or outcome was good? If those criteria are known, the people executing can self-check their results against them, rather than waiting for a boss to approve.

All of these mechanisms, policies, priorities, escalation criteria, success metrics, act as embedded supervision. They carry the intent and judgment of leadership into each decision without requiring the leader to stand over everyone's shoulder. The result is a more scalable model of control. A single leader's influence can permeate dozens of decisions via a well-crafted policy or a clearly communicated priority, whereas previously that leader might have only been able to directly review a handful of decisions per day. Importantly, this does not mean a loss of control where it matters. Leaders still identify the pressure points and make sure that critical or novel issues get human attention. But they aren't wasting their limited bandwidth on rubber-stamping every minor choice. Instead, they spend that time and energy shaping the system that shapes the choices. This is a fundamental reorientation of responsibility: from doers of second-order tasks (reviewing, approving, correcting) to designers of first-order conditions (intent, rules, context).

The daily life of a manager changes under this model. Imagine two department heads: One spends the morning checking over each team member's work output, correcting small errors, and giving ad hoc approvals, their influence is direct but narrow. The other spends the morning analyzing why certain errors keep happening, and then updating a training manual or decision guideline to eliminate that error in the future, their influence is indirect but broad. The first manager is practicing supervision; the second is practicing intent-based guidance. Over time, the second manager's department will likely see fewer repetitive mistakes and will function more autonomously, because the manager's intent has been baked into how the team operates. This approach also tends to be more satisfying for both parties: team members feel more trusted and empowered, and managers feel less like babysitters and more like true leaders. It aligns with the often-stated desire of leaders to "focus on strategic issues" rather than getting bogged down in the weeds. By shaping decisions instead of checking work, leaders elevate their perspective to the level of systems and principles. They still exercise control, arguably even more effective control, but through foresight and guidance rather than after-the-fact inspection.

WHAT REMAINS UNDER HUMAN SUPERVISION

In highlighting the shift toward intent and upstream guidance, it is critical to dispel a possible misinterpretation: that human supervision will disappear entirely. It will not, and should not. Even in the most advanced, intent-driven organizations, there are situations that inevitably demand direct human oversight. The evolution from supervision to intent is about using supervision more selectively and effectively, not eliminating it wholesale. So, what kinds of work or decisions remain under human supervision, and why do they still matter? Generally, these are the cases characterized by novelty, high stakes, ethical complexity, or system failures, in short, situations where rules and intent might not yet suffice and where human judgment is irreplaceable.

First, novel or unique scenarios often require human oversight because no pre-defined intent or policy covers them. If something entirely new comes up, say a one-of-a-kind customer request, an unprecedented technical glitch, or a crisis that wasn't anticipated, there is no substitute for a human supervisor stepping in, understanding the context, and charting a course. By definition, intent can only guide predictable situations; when unpredictability reigns, direct supervision is the fail-safe.

Second, high-risk decisions are frequently kept under human supervision as an extra safeguard. For example, in medicine, no matter how clear the protocols (intent) are, a complex surgery or a life-and-death triage decision will involve experienced surgeons or physicians actively overseeing and adjusting in real time. In finance, a multi-billion-dollar investment decision will go through intense human scrutiny even if junior analysts have prepared the ground. The higher the stakes, the more we desire a responsible person's eyes on the matter, to double-check and take accountability at the moment of action. Automation and guidelines might handle routine trades or diagnoses, but a human is still expected to oversee the exceptional, weighty moments where the cost of a mistake is enormous.

Third, areas involving ethical judgment or interpretation of values remain under human watch. No policy can perfectly encode a company's values or society's expectations for every situation. Responsible oversight is needed when decisions have moral or legal dimensions that require nuance. For instance, consider content moderation on an online platform: algorithms (guided by intent in the form of rules) do a lot of the work, but human supervisors are needed for borderline cases and to refine those rules over time. Or consider a company navigating a legal gray area in a new market; leaders will personally supervise those choices because they involve interpreting regulations and ethical norms that can't be reduced to a formula. Human judgment, with all its depth, comes to the forefront here. Indeed, regulators themselves often mandate human oversight for certain automated systems: the European Union's AI Act of 2024, for example, requires that high-risk AI decisions be "effectively overseen by natural persons" during operation. The message is clear, some domains are too sensitive to leave entirely to automated intent without a human in the loop.

Fourth, breakdowns of assumptions call supervisory attention. If the organization has been operating under certain assumptions (e.g. a model of customer behavior, a stable market condition, a calibrated machine setting) and those assumptions suddenly no longer hold, humans must intervene. This is the classic "exception handling" role of management. You might set up a brilliantly intent-driven process for, say, supply chain ordering, but if a once-in-a-century storm knocks out a supplier, someone with authority needs to step in and improvise. The system can be designed to escalate such issues precisely because we expect human judgment to take over in those moments. In other words, when the environment throws something truly outside the model, supervision comes back to the forefront as the mechanism of responsibility.

It's worth noting that mentorship and development of people is a facet of supervision that remains vital. Guiding intent is great for established processes and skilled teams, but less experienced employees often need closer coaching initially. A supervisor may deliberately "oversupervise" a newcomer not to control them forever, but to accelerate their learning and ensure they grasp the intent and standards of the organization. This kind of hands-on teaching is an age-old role of supervisors (think of apprentices and master craftsmen), and it doesn't vanish in an AI-enabled, process-refined world. If anything, the value of such mentorship increases as routine supervision decreases, because what supervision remains is more focused on building capability and handling the tough stuff. When mundane checks are automated or eliminated, human supervisors can concentrate on higher-level support: refining judgment, conveying tacit knowledge, and instilling the organization's culture and values in their teams.

Human supervision remains indispensable for the unforeseeable, the high-impact, the value-laden, and the developmental. As we shift many aspects of control upstream to intent, the instances where we do choose to apply direct oversight become more deliberate. We save our supervisory capacity for where it truly counts. And as a consequence, the value of human supervision actually increases as it becomes more selective. When a leader steps in now, it's for something truly deserving of their attention, not just a routine sign-off. This selective presence reinforces why leadership and experience still matter greatly, they are applied where algorithmic or procedural control reaches its limits. The challenge for organizations is to clearly identify these domains and preserve robust supervisory practices around them, even as other areas are delegated or automated. In doing so, they ensure that "intent-driven" does not mean "hands-off entirely," but rather hands-on when and where it truly matters.

WHY INTENT DOES NOT DILUTE ACCOUNTABILITY

A common concern in shifting from supervision to intent is whether accountability will be weakened. If managers are no longer in every loop, how do we make sure people remain answerable for what happens? Does giving people more autonomy to act on intent let them off the hook? The answer is that intent changes how accountability is enforced, not whether it is. In fact, when done right, an intent-driven approach can make accountability more transparent and harder to evade. It recenters accountability on outcomes and clear standards, rather than on process compliance or personal presence.

In a traditional supervised model, it's possible for accountability to become muddled. A subordinate might say, "I did exactly what my boss said to do," implying that the boss is accountable for the instructions, while the boss might say, "Well, I followed the procedure that was in place," implying the system or someone else is accountable for the results. This is how responsibility can diffuse in heavily layered oversight environments, a phenomenon sometimes called accountability distortion. People focus on showing that they followed the rules or orders (to prove they're not personally at fault), rather than focusing on whether the outcome was actually achieved. The illusion of safety through compliance can undermine real responsibility. As one leadership commentator put it starkly, "Compliance gets short-term results. It ensures the job is done while someone is watching. But the moment supervision fades, so too does the urgency," leading people to do just enough to avoid scrutiny. In other words, if accountability is defined only as adhering to your supervisor's immediate requirements, then when that supervision is absent, accountability can slip away. The individual might not feel personally responsible for the result, only for following instructions.

Intent-based control, by contrast, makes accountability more direct. Because goals and boundaries are explicitly stated, everyone involved knows what the collective aim is and what limits cannot be crossed. Each actor is empowered to make decisions within that space, which also means they own the consequences of those decisions. If you are given the intent ("achieve X outcome under Y conditions") and adequate resources and support, then delivering that outcome is part of your accountability. You can't say "well, no one told me exactly how, so it's not my fault", the intent did tell you what was expected. Likewise, leaders who set the intent are accountable for the clarity and wisdom of that intent. If a team keeps making decisions that technically align with the stated intent but lead to a bad outcome, it suggests the intent was flawed or ambiguous. That is on the leader to fix. Thus, instead of accountability being about did you do the task as you were told, it becomes did we achieve the outcome we set out to, and if not, why? There is far less room to hide behind procedure. Each deviation or failure prompts a look at either execution or guidance: did someone make a poor choice, or was our guidance insufficient? Either way, the conversation is anchored in results and explicit expectations, not in whether a checklist was followed.

It's important to stress that none of this erases hierarchy or authority. Shifting to intent does not mean leaders abdicate their authority; it means they exercise it in a different manner. A leader still has the authority to set direction, to make final calls on difficult trade-offs, and to intervene in crises. In fact, being explicit about intent reinforces a leader's authority in a healthy way, their influence is baked into the operating criteria of the organization. Subordinates acting on intent are effectively carrying out their leader's mandate, just without the leader dictating each move. Moreover, authority is preserved in the ability (and responsibility) to step in when intent isn't enough, as we discussed in the previous section. If a situation exceeds the scope of the given intent, it escalates to higher authority. So hierarchy doesn't vanish; it becomes a backstop rather than a throttle on every decision.

What does change is how we measure and discuss accountability. In an intent-driven model, accountability is measured by outcomes against the stated intent. Did the project meet the goals within the constraints? Did the team uphold the priorities and values laid out? If not, those are clear grounds for accountability discussions. Contrast this with a compliance model where someone might say "I followed all the steps I was given" as a defense, which could be technically true and yet the outcome failed.

Under intent, following steps is not a get-out-of-jail card; delivering on intent is the metric. Paradoxically, this can feel stricter in some ways: it's outcome accountability in a purer form. But it's also fairer. People aren't punished for coloring outside the lines if the outcome is good and within the intent; conversely, they can't claim virtue for coloring inside the lines if the outcome is poor. The conversation shifts to whether the intent was met, which is a more meaningful conversation for the organization's success.

An added benefit is that accountability becomes more shared and transparent. When a whole team understands the intent behind their mission, they collectively watch that they are staying on course. Peer accountability often increases, because everyone knows what the target is. It's not just the boss catching errors; teammates will call out, "Hey, if we do that, we'll violate the constraint we were given," or "We're drifting from the priority, let's refocus," because the intent is clear to all, not a private checklist in a manager's drawer. In this sense, explicit intent creates a kind of distributed oversight, which complements formal accountability structures. It turns "being accountable" into part of the culture, rather than just a top-down enforcement. People choose accountability, as some leadership writers say, rather than only feeling accountable when watched. They internalize the goals and standards, which is the strongest form of responsibility.

To be sure, the transition to this model must be managed carefully. If an organization simply stopped supervising without providing clear intent and building a culture of trust and competence, accountability would suffer; chaos would likely ensue. But that would be a mistake of execution, not a flaw in the concept. The proper sequence, as explored in earlier chapters, is to ensure the building blocks, clarity of purpose, reliable execution mechanisms, transparency of decisions, are in place so that intent can actually govern day-to-day actions. When those conditions are met, moving from supervision to intent does not weaken control or accountability. It shifts them to a form that is more scalable and ultimately more aligned with results. People are still accountable to their superiors and stakeholders; superiors are still accountable to set the right direction and to intervene when needed. What changes is that everyone gains a bit more ownership of doing the right thing even when nobody is looking over their shoulder. And ironically, when nobody needs to be looking over your shoulder because you're guided by intent, it often means accountability has truly taken hold within the individual. They are doing the right thing because it's the right thing, not just because a supervisor will catch them if they don't. In that sense, intent-driven responsibility is not a dilution of accountability at all, it is a deepening of it, woven into the fabric of how work gets done.

WHAT THIS CHAPTER ESTABLISHES

We have seen in this chapter that supervision, the direct, eyes-on oversight of human work, was a logical and effective response to the challenges of a human-driven execution model. When every task depended on individual effort and judgment, organizations quite sensibly put managers and supervisors in place to observe and guide that effort in real time. Supervision made outcomes more consistent and ensured accountability by putting responsibility at the point of action. This was neither foolish nor archaic; it was essential for much of modern organizational history, and it succeeded in enabling large enterprises despite human limitations.

At the same time, we have examined why that supervision-centric model strains under scale and complexity. It does not linearly extend to a world of massive operations and rapid, intricate decision flows, at least not without unacceptable costs in speed and resources. Human attention is the limiting factor. Thus, as execution becomes more reliable (through better processes, training, or technology) and as coordination costs begin to dominate, the locus of effective control shifts. We introduced intent as a higher-leverage way to guide execution: one that applies control at the level of direction and boundaries rather than through continuous monitoring. Intent allows a leader's influence to scale farther, by shaping many decisions with a single clear articulation of purpose and constraints. It is, in effect, a refinement of organizational design: a move from hands-on control to heads-up control.

Importantly, this shift from supervision to intent builds on the earlier shifts we've discussed (from process to policy, from roles to decisions), it doesn't discard the past so much as reallocate its wisdom. We are not talking about flattening organizations into free-for-alls or discarding hierarchy and expertise. Rather, we are describing an evolution of control. Policies (Chapter 7) set broad boundaries, decision-centric thinking (Chapter 8) targets where oversight is needed, and now intent (Chapter 9) becomes the primary mechanism for guiding execution at scale. Together, these changes depict a model where control is exercised more upstream and more selectively. Supervision doesn't vanish; it specializes. Leaders cease to be omnipresent inspectors and instead become enablers of autonomous execution and guardians of intent.

By understanding this progression, we dispel the false choice between total control and total chaos. There is a middle path where an organization can be highly controlled and highly empowering: controlled through clarity of intent, empowering by granting freedom within that clarity. Accountability remains every bit as strong, arguably stronger, in such a model, it's just measured against outcomes and explicit expectations rather than against adherence to micromanagement. The role of human judgment remains vital, but it is deployed where it adds the most value, not squandered on trivial checks.

This chapter establishes that the move from supervision to intent is a foundational structural shift for modern organizations. It's a response to new realities: larger scales, faster operations, and more consistent execution technology. It asks leaders to exercise responsibility in a new way, one that can feel less immediate (because you're not right there at the elbow of your team member) but is ultimately more potent. Of course, making this shift is not without risk or difficulty, and it must be done under the right conditions. The next chapter will address those very dangers: what happens if we try to pull back supervision too early or without the proper supports in place. As we will see, succeeding with an intent-driven approach requires preconditions like clear communication, trust, competency, and reliable systems. Abandoning old controls without those in place can backfire, leading some to conclude that "intent doesn't work" when in fact the groundwork wasn't ready.

The evolution from supervision to intent recenters how we think about control and responsibility. It reminds us that the essence of leadership is not about hovering over each task, but about setting the direction so that tasks align naturally. It reaffirms that accountability is not an artefact of having a supervisor watching; it is an intrinsic duty tied to outcomes and values. And it reassures us that even as we embrace new modes of control, we carry forward the core rationale of the old, the commitment that outcomes must be seen to and cared for. We are not eliminating supervision; we are elevating it to a higher plane of influence. In that sense, intent is supervision by other means: a way for responsible people to fulfill their obligations in a world where direct oversight simply cannot reach everywhere. The responsibility remains, firmly and resolutely, only now, it acts through intent.

10

THE RISKS OF MOVING TOO EARLY

Leaders in modern organizations often reach a point of clarity about how their operating model needs to evolve. They see the promise of a new approach, perhaps shifting from detailed process manuals to broad policy guidelines, from rigid role definitions to flexible decision-making, or from close supervision to guided intent. Understanding these structural shifts is an important insight. However, recognizing what should change is not the same as being ready to make that change. Structural insight does not automatically confer structural readiness. In practice, many well-intentioned leaders act on a correct vision for the future before their organization can safely absorb it. The result is not bold transformation but a subtle form of failure: the change that comes too early and quietly misses the mark.

This chapter focuses on those quiet failures. Moving too early rarely produces a dramatic collapse that sounds alarms. Instead, it often leads to a gradual drift, a period of ambiguity, inconsistency, or confusion that may only later be recognized as a failure. Because the breakdown isn't explosive, it can be hard to diagnose. Was the new model flawed? Or was the timing off? In many cases, the answer is the latter. The direction of the change can be perfectly sound, yet it's premature execution undermines its very goals.

Importantly, moving too early is not a sin of malice or incompetence. It's often born of enthusiasm and a sincere belief in a better way. A leadership team gains insight from a promising pilot, an industry trend, or an internal analysis and decides to implement structural changes right away. What they sometimes overlook is that preparation and sequence matter as much as direction. Each structural shift we have explored, from process to policy, roles to decisions, and supervision to intent, depends on enabling conditions to succeed. Clarity of intent, explicit and well-understood policies, transparent decision rights, and reliably consistent execution are the underpinnings of the new model. If those foundations are not yet in place, attempting the shift is like trying to operate a machine without all the parts assembled.

The sections that follow examine the risks of implementing these shifts before an organization is ready. In each case, we will see how a change that is sound in principle can backfire if done without the necessary groundwork. The aim is not to discourage change but to highlight the discipline of timing, to show that restraint and careful sequencing are themselves forms of organizational competence.

WHEN POLICY REPLACES PROCESS BEFORE INTENT IS CLEAR

One of the major shifts envisioned in a modern operating model is moving from detailed processes to broader policies. In theory, this is empowering: teams are given guiding principles ("policies") instead of step-by-step instructions ("processes"). They gain flexibility to choose how to meet objectives, rather than following a rigid script. But if this change is made before everyone shares a clear understanding of intent, it can create ambiguity rather than autonomy.

Processes, for all their rigidity, carry built-in clarity. A well-defined process spells out exactly what to do next and who must do it. When leadership decides to remove a process or lighten procedural controls in favor of high-level policy, they must ensure that the intent behind the work is explicitly understood by all. If not, each individual or team may interpret the policy in their own way. Without concrete steps to follow or a crystal-clear objective, employees might substitute their personal judgment in inconsistent ways.

For example, consider a company that historically required multiple approval steps for any new project. Suppose leadership, eager to increase speed, announces that those approval processes are no longer mandatory, replaced by a simple policy: "Use good judgment and move fast." If this policy remains vague, if "good judgment" isn't contextualized with boundaries or the organization's strategic intent, employees will have varying interpretations. One team might charge ahead with a high-risk project believing they are empowered to do so, while another hesitates on a moderate decision because they are unsure what "move fast" truly permits. In the absence of the old process, people create their own rules. The result can be fragmentation and risk. Decisions start to diverge in quality and alignment. Some corners may be cut under the banner of speed. Without a shared intent, the organization loses the very predictability and coherence that the prior process ensured.

Eventually, consequences appear. Perhaps an avoidable error or compliance lapse occurs under the looser regime. Management, alarmed by rising variation or a near-miss, steps back in to assert control. Ironically, the response to a failed early move toward policy is often to reintroduce process, sometimes in an even heavier form than before. In our example, after a project goes awry, leadership might restore the approval chain and add extra checkpoints "to be safe." What was meant to liberate teams ends up reinforcing bureaucracy. The lesson many take away is that the policy-driven approach "doesn't work," when in truth the organization had not set the necessary clarity of intent upfront.

As we discussed earlier in this book, a policy can only substitute for process when it is backed by unambiguous understanding. Teams need to know the non-negotiable goals, the constraints they must respect, and the values that should guide their choices. If "use your judgment" is meant sincerely, it must be accompanied by shared context: what risks are absolutely unacceptable, what outcomes matter most, and where to seek guidance when unsure. Absent that scaffolding, a sudden switch to policy is just ambiguity packaged as empowerment. It feels like freedom until something goes wrong.

In short, policy-driven autonomy is powerful only after intent is clear and broadly internalized. Moving to that model too early, before people know, with confidence, the purpose and parameters of their freedom, sets the stage for inconsistency and error. The risk here is subtle: for a time, everything might appear fine, with teams enjoying newfound latitude. But without common guardrails, that initial calm is fragile. It takes just one team's well-meaning misstep to reveal that "empowerment" was really confusion in disguise. True empowerment requires clarity; without it, replacing process with policy simply trades one type of constraint for another, more dangerous one: uncertainty.

UNBUNDLING ROLES WITHOUT REDESIGNING ACCOUNTABILITY

Another envisioned structural shift is moving from rigid roles to a more fluid, decision-centric organization. In a traditional model, work is organized by roles and departments, each with defined responsibilities. The new model suggests we focus more on decisions and outcomes rather than fixed titles. In practice, this might mean loosening role boundaries, encouraging people to take on decisions outside their old silos, or flattening layers of management. Done right and at the right time, this can make an organization more agile and collaborative. Done too early or carelessly, it can create dangerous gaps in accountability.

Roles have long been the backbone of accountability. When everyone knows "Alice is the finance manager who approves budgets" or "Bob is the project lead who coordinates execution," there is clarity about who owns what. If we suddenly "unbundle" these roles in the name of flexibility, saying, for instance, that formal titles matter less now and teams can self-organize around decisions, we must simultaneously redefine who is accountable for each decision and outcome. Failure to do so means that when the next budget needs approval or a project hits a snag, it may be unclear who is responsible. Work can stall or fall between the cracks because the old structure has been removed and no new structure catches the task.

Consider an organization that tells its department heads that, moving forward, decisions should be made by those with the most knowledge of the issue rather than by hierarchy. This is a well-meaning step to empower subject-matter experts. But if those department heads' roles are effectively diminished without a new clarity on decision ownership, the experts themselves might be unsure: "Am I truly authorized to make this call? If I decide and it goes wrong, who is accountable, me or the former decision-maker?" In the vacuum of role definitions, two reactions often emerge. The bolder individuals may charge ahead and make unilateral decisions, sometimes stepping on toes or duplicating effort. The more cautious individuals do the opposite, they hesitate, or they seek approval informally because they aren't confident they have the authority. In both cases, the lack of explicit accountability undermines the promise of a role-light, decision-centric model.

Real-world experiments bear this out. One high-profile example came from a company that eliminated most manager titles in favor of self-managing teams. The theory was that without rigid roles, employees would naturally take ownership of the decisions in their domain. What happened instead was a period of confusion and accountability drift. Some former managers struggled to find how to contribute and inadvertently abdicated oversight that was still needed. Meanwhile, some team members assumed someone else was handling a critical decision, and when that proved false, deadlines were missed. Former employees from that experiment noted that without clear accountability, it became "hard to hold staff accountable or get people to take responsibility for things that need to be done". Even highly motivated people were unsure about their authority, and many became either indecisive or overstepped boundaries. In the absence of defined leadership roles, a significant number of people simply weren't sure what to focus on or felt less compelled to deliver on nebulous responsibilities, as one observer quipped, if you suddenly tell a group they have no managers, a few will thrive on the freedom but "the other 8 or 9 will either slack off, not know what they're supposed to focus on, or [encounter] other problems".

These outcomes are not arguments against distributing decision-making. They are reminders that accountability must be reattached when roles are unbundled. Letting go of titles and org charts alone doesn't automatically distribute knowledge of who answers for what. If anything, it can mask accountability: problems emerge but no one is clearly on the hook, because the old owner isn't in charge and the new owner was never named. An organization that moves to a decision-centric ethos prematurely will likely experience a spike in "I thought you were handling that" scenarios. Work fragments; decisions escalate upwards again or simply languish. People start creating ad-hoc workarounds to fill the void, maybe an unofficial committee forms to approve things that a single role used to handle, or individuals quietly revert to seeking permission even though they're told they're empowered.

All this can erode trust and effectiveness. Co-workers begin to doubt the reliability of commitments ("Who exactly is making sure this gets done now?") and risk-aversion grows. In some cases, rather than feeling liberated by the change, employees feel exposed. They fear making decisions that were once a manager's domain, because if something goes wrong, it's not clear whether they will be supported or blamed.

Removing role boundaries without deliberately redesigning accountability is like removing the pillars of a structure without adding new support beams. The structure doesn't collapse outright on day one, but stress cracks appear. To avoid this, any shift away from roles needs an accompanying, explicit mapping of decision ownership: a clear answer to "Who is responsible for this decision or outcome now?" Until that mapping is firmly in place and accepted, a premature liberation from roles will create more hesitation and handoffs than bold decisions. It will replace one form of coordination (the org chart) with chaos, which is not an improvement at all.

REDUCING SUPERVISION BEFORE EXECUTION IS RELIABLE

A third structural shift we have explored is moving from supervision to intent, that is, reducing day-to-day oversight as teams or systems become capable of self-directing their work under a clear intent. The promise here is high leverage: leaders articulate what needs to be achieved ("the intent") and trust teams, possibly augmented by advanced systems, to execute without constant check-ins. This can unlock creativity and speed, as it relieves the organization from micromanagement. But if supervision is pulled back before the execution can truly stand on its own, the likely result is inconsistency, hidden errors, and a quick erosion of trust.

Supervision has always been a safety net. Managers and supervisors exist in traditional structures to catch mistakes, to ensure quality, and to intervene when something deviates from plan. If we imagine an environment where execution (whether by human teams or AI-assisted processes) is perfectly reliable, then heavy supervision indeed becomes unnecessary. However, in reality, reliability is rarely absolute from day one. New processes, new algorithms, or newly autonomous teams need time to prove they can consistently meet their targets and handle the unexpected. If leadership enthusiastically declares, "We won't be looking over your shoulders anymore, just follow the intent and use your best judgment," they must be certain that the systems and people are ready to uphold standards independently.

When that certainty is absent, cutting oversight is a gamble. In the early phase of any change, there will be variance: a team still learning to self-regulate might miss subtle cues that a seasoned supervisor would catch, or an automated system might not yet handle every corner case. By stepping back too soon, leaders create a condition where small problems go unnoticed and uncorrected longer than they would have under closer watch. Minor deviations compound. Because everyone has been told they are trusted to execute, they may also assume they shouldn't "bother" management with small concerns, until those concerns grow big. The initial period after reducing supervision can actually feel liberating and quiet; it's only later that unseen issues surface, sometimes suddenly, revealing that the consistency was never truly there.

The effect on trust can be dramatic. Imagine a manufacturing operation where a supervisor used to inspect output at the end of each day. Eager to adopt a modern, intent-driven approach, management tells the team that inspections will be infrequent now because they trust the process and the workers' diligence. For a few weeks, all seems well, production runs continue without obvious incident. But in the third month, a batch of defective products reaches customers before anyone catches the quality issue. Upon investigation, it turns out the team had not fully understood a subtle calibration requirement; previously, the supervisor's daily checks caught this, but once oversight was removed, the calibration drifted. Now customers are affected. The leadership's immediate reaction is alarm and a feeling of broken trust: perhaps they conclude that "the team wasn't ready" or even that "this self-regulation idea doesn't work." They reinstall daily inspections or add even more checkpoints than before, and the experiment in reduced supervision ends with disappointment. Notably, the team might feel just as betrayed, they were told they were trusted, only to be treated as untrustworthy after a failure that they had little support to prevent.

Such scenarios highlight that trust in execution must be earned and demonstrated through consistency. It cannot simply be declared by fiat. When an organization moves to lighten supervision, it should be because prior evidence suggests the process is stable and that workers or systems have the capability to monitor themselves effectively. If that evidence is not solid, reducing oversight is premature. As some change experts have publicly observed, when a drastic new way of working is forced too quickly, the organization's "immune system" may kick in. People revert to old habits or informal checks to compensate, or they panic and overcorrect when the inevitable variances appear. In the face of errors, leaders too can become reactionary, concluding that the entire approach of guiding by intent has failed, when really it was a timing issue or a scope issue (perhaps it would have worked on a smaller scale or with more training).

The risk of moving too early to an unsupervised model is not just the immediate operational mistake, but the loss of confidence it triggers. A single incident can cast a long shadow. Executives who were once excited about empowering their teams may lose faith in the concept after a premature stumble. They may feel burned, and teams, in turn, become more cautious. In the worst case, an early lapse creates a narrative that "self-management (or automation, or whatever new approach) can't be trusted," hampering the organization's appetite for future innovations. Thus, pulling back supervision too soon doesn't merely revert the organization to its old state; it often pushes it further back into a state of distrust, with even tighter controls than before.

The clear message is that reducing supervision is viable only after reliability is proven. This might mean staged trials, shadow periods where oversight continues quietly in the background, or incremental hand-offs of responsibility. The organization must see a track record of the intent-driven model working safely. Until then, stepping away entirely is an unnecessary risk. Just as a mentor gradually gives a protégé more autonomy as they show competence, organizations should phase in autonomy in proportion to demonstrated capability. Doing otherwise is not boldness; it's a lack of due diligence.

WHY EARLY FAILURES ARE INTERPRETED AS PROOF

When an organization attempts a structural change and encounters failure early on, a common and human reaction is to treat that failure as definitive proof that the new model is fundamentally flawed. This tendency has less to do with the objective merits of the new approach and more to do with psychology and the comfort of the status quo. Early failures loom large in the collective memory of an organization. If not carefully examined, they can harden into cautionary tales that reinforce commitment to the old ways.

Think of a team that experiments with a new decision-making process, only to have a visible project derail. The post-mortem discussion rarely attributes the failure to "we tried this too soon" or "we didn't have the right supports in place yet." Instead, one is likely to hear things like, "See, that new approach doesn't work for us. We gave people more leeway and look what happened." The failure is interpreted as evidence that the idea itself, be it policy over process, flattened roles, or reduced oversight, was a mistake. People understandably point to tangible negative outcomes, concluding that the traditional model would have avoided them. It's a form of confirmation bias: if some employees or managers were skeptical of the change to begin with, the first sign of trouble confirms their suspicions that the old model was safer.

Organizations also have long institutional memories when it comes to failure. A failed early attempt can become a folk warning passed down in meetings and hallway conversations: "Remember when we tried that new system and it backfired? Let's not go there again." Such memories often lack the nuance of why something failed, it becomes a simple tale that "we tried X and it didn't work." In truth, the new system or structure may have been perfectly viable under different conditions, but that context is lost. What remains is a general wariness. Leaders who championed the change may lose credibility, and those who were comfortable with the old ways gain validation.

This dynamic reinforces the existing model. After an early failure, there is a natural retreat to familiar territory. If a foray into policy-driven governance leads to a stumble, the organization responds by doubling down on process and micromanagement: never again, until further notice. If a step toward self-managed teams results in confusion, upper management might swiftly reinstall layers of approval and oversight, perhaps more than existed originally, "just to be sure." Each retreat hardens the organization's immune system against future change. The next time someone proposes a structural shift, they face not only the inherent challenges of the new approach but also the ghosts of past failures: "We've already been down that road."

The risk here is that premature experiments, if not framed and learned from properly, make later, better-timed transitions harder. An early failure can reduce the appetite for innovation for years. In some organizations, one will often hear references to an attempt long ago as the definitive reason a new idea is off the table: "We tried a remote work policy in 2019 and productivity dropped; we won't try that again," or "We once eliminated an approval layer and it caused a customer issue; we need that buffer." These statements often ignore how conditions might have changed or how execution could be improved. They serve as convenient stop signs, preventing even well-prepared initiatives from gaining support.

It is important to note that this phenomenon is not about irrational or stubborn people clinging to the past for its own sake. It is about trust and credibility. A failure, especially one that impacts real outcomes, chips away at trust in the new idea and in those who advocated it. People with real responsibility for outcomes (the kind of readers of this book) do not easily forget a change that hurt their ability to deliver safely. They become more conservative in the aftermath. In their eyes, being skeptical is being responsible; they've been burned once, and their job is to not let it happen again.

Thus, the cost of moving too early is not only the immediate stumble but also a deeper entrenchment of the old model. Early failures give ammunition to voices that say "if it isn't broken, don't fix it," or more pointedly, "you broke it by trying to fix it." It can take a long time to rebuild momentum and trust after such an event. That is why careful sequencing and pilot testing (without overhyping them) are so crucial, to avoid turning the organization against the very changes it will eventually need to embrace. A failed early attempt is often worse than no attempt in the short run, because it creates organizational scar tissue. Ironically, a mistimed, premature change can push an organization into a posture of delaying all change due to lost confidence. In that sense, moving too early can indirectly cause the opposite problem: it can make an organization so risk-averse that it ends up moving too late when change truly becomes critical.

THE COST OF LOSING PROTECTIVE MECHANISMS TOO SOON

Underlying all these points is a fundamental truth: the current organizational model, with its processes, defined roles, and supervision layers, exists for reasons. Chief among those reasons is protection. These mechanisms have been the way organizations ensure quality, compliance, ethical standards, and consistency at scale. If you remove those protective mechanisms before new ones are in place, you are essentially disarming the organization's defenses prematurely.

Consider compliance and legal safeguards. In a heavily regulated industry like healthcare or finance, there are good reasons for all the paperwork, approvals, and double-checks. Those routines act as checks against violating laws or ethical standards. If a company in such an industry decides to drastically streamline or drop many of these processes overnight because it has a new policy-based approach in mind, it may inadvertently open itself up to serious risk. Without equivalent controls embedded in the new way of working, compliance exposure grows. The organization might find later that it failed an audit or breached a regulation because the manual sign-offs it thought were mere busywork were actually catching errors or ensuring accountability.

The same goes for ethical drift. Process and oversight help uphold an organization's values by providing moments where someone asks, "Is this the right thing to do?" When moving to a faster, leaner model, if too many of those moments are removed without alternative ways to enforce values, people under pressure might make expedient choices that conflict with the organization's principles. This isn't because they are unethical, but because the guardrails that reminded everyone of standards were lifted. History is unfortunately full of scandals in organizations where controls were relaxed or bypassed in the rush to achieve results; only afterward does it become clear that those controls served an ethical purpose.

Operational consistency is another protective aspect. A standard process ensures that no matter which individual is doing the work, the approach and outcomes are within a predictable range. It prevents the scenario where one team's output is excellent and another's is flawed because they took wildly different approaches. If processes are cast aside wholesale, a company could see wide variability in operations. One location or unit might start doing things very differently from another, believing they are following policy or intent, but in reality diverging. This can damage quality and brand reputation, and even lead to safety issues if, say, different manufacturing plants start applying different safety practices in the name of local empowerment.

And then there's reputational damage. Customers, partners, and even employees themselves come to trust an organization based on steady performance and reliable standards. A controlled process, a clear owner, a supervisor's sign-off, all these reassure stakeholders that things are under control. If an organization peels these away prematurely and something goes wrong, the external impression can be harsh: "They were negligent" or "They moved too fast and broke things." Rebuilding an external reputation after a public failure is much harder than maintaining it in the first place. Internally, employees might feel embarrassed or let down that leadership exposed them to risks they weren't ready to manage.

Earlier in this book, we examined how processes can act as protective scaffolding, not arbitrary red tape, but deliberate defenses against known failure modes. The lesson to carry forward is not that these structures must remain forever, but that when we remove them, we must have new protections actively working from day one. No one decommissions the fire department just because smoke detectors were installed. Similarly, oversight shouldn't be dropped just because a high-level policy has been drafted; leadership must ensure the policy is ingrained and that people or machines can execute it with the same reliability the old checkpoints provided.

Moving too early often means there is a gap in protection. Everyone hopes nothing will slip through in that gap, and indeed, for a time, things might seem fine. But those gaps are often invisible until, as the saying goes, the tide goes out. It might take months or a crisis to reveal that some compliance requirement was overlooked or that a critical task had no true owner. At that point, the damage is done. The tragedy is that such damage was preventable with more patience: by holding onto certain safeguards until the new system demonstrably covered those needs.

None of this implies that the new model cannot be as safe or safer than the old. It certainly can, often the goal of new structures is to maintain or improve safety and performance while gaining flexibility or speed. But timing is everything. If, for example, automated checks are supposed to replace human oversight, those checks need to be tested and proven at small scale while the human oversight still runs in parallel. If decentralized decision-making is to replace centralized control, there must be training, clear ethical guidelines, and monitoring in place during the transition. Prematurely assuming everything will work perfectly is a form of organizational hubris that exposes the firm to outsized risks.

Ultimately, the cost of losing protective mechanisms too soon is getting the downside (exposure to harm) without the upside (the fully realized new efficiencies or agility). It is a lose-lose proposition: the organization becomes less safe, but not actually much faster or more innovative because the backtracking will soon begin. The better path is a balanced one, retaining protections until they are genuinely redundant, not casting them aside in the excitement of a vision.

WHAT THIS CHAPTER ESTABLISHES

Moving too early to a new structural model carries very real risks, not because the new model is wrong, but because sequence matters. We have seen that replacing process with policy only works when intent is crystal clear; unbundling roles only works when accountability is explicitly redefined; reducing supervision only works when consistent execution has been proven. None of these shifts are simple flips of a switch. They are changes in how the organization provides clarity, alignment, and safety. If those provisions are not yet fully developed in the new form, the old form (process manuals, org charts, oversight) will painfully prove its continued relevance by the void it leaves behind.

This chapter is not a counsel of paralysis or an argument to cling to the past indefinitely. It is, however, an argument for discipline in timing. There is a competence in knowing not just how to design a new structure, but when to implement it. Restraint can be a virtue in leadership; it signals respect for the complexity of the organization and the seriousness of the outcomes at stake. By approaching each structural shift with patience, by saying, in effect, "We will not remove this support until we are confident we have another one in place", leaders protect both their people and the credibility of the change itself. It takes maturity to hold back when the path is conceptually clear but the groundwork is not yet sufficient. That maturity distinguishes sustainable transformation from a gamble.

When moves are made in the right sequence, change can happen with minimal disruption and without undermining trust. Early successes (even small ones) then build confidence and momentum, rather than early failures breeding cynicism. The goal is not to avoid change, nor to delay it out of fear, but to execute it with a sense of order, aligning it with the organization's readiness. In doing so, leaders increase the likelihood that when the transition happens, it sticks for the right reasons instead of backfiring.

Yet focusing only on the dangers of moving too early presents just one side of the challenge. If haste can undermine progress, so can hesitation. An organization that clings too long to familiar models risks a different set of failures, ones that creep in gradually under the guise of stability. Even a well-functioning model, held beyond its time, can quietly accumulate fragility as circumstances change around it. Thus, understanding the perils of premature change naturally leads us to consider its opposite: the risk of moving too late.

Effective leadership is ultimately about striking this balance, knowing when to hold steady and when to press forward. By internalizing the lessons of moving too early, we prepare ourselves to navigate the equally significant dangers of moving too late. Only by keeping both perspectives in view can we find the prudent path that steers our organizations through structural change safely, responsibly, and successfully.

11

THE RISKS OF MOVING TOO LATE

Stability in an organization is often equated with safety. Processes run as expected, responsibilities are clear, and outcomes remain predictable. However, stability and true safety are not the same thing. A structure that has proven itself over time can quietly accumulate hidden fragilities when the environment around it changes. Often, what changes is not the organization itself but the context in which it operates: customer expectations shift, technologies evolve, competitors introduce new ways of working. A system designed for yesterday's conditions can become ill-suited for today's. For example, a workflow that was seamless when everyone worked on-site may start to falter in a remote or hybrid work setting; extra coordination and clarifications become necessary, though it's not immediately evident in top-line results. Changes in the business environment, whether driven by technology, customer behavior, or external events, can gradually turn a well-tuned process into a clumsy one. In the earlier chapters, we established why the existing operating model earned its place, it provided reliability in a world of human limits and built a foundation of mastery and trust. Yet those very strengths can become weaknesses if conditions shift. The success of the old model can create a false sense of security, a reluctance to question what has long worked. What was once a source of confidence can turn into a source of complacency.

One sign of this creeping fragility is an increase in the effort required to achieve the same outcomes. The organization might still meet its goals, but it does so with more meetings, more exceptions, and greater reliance on workarounds. Small inefficiencies that were once negligible begin to multiply. Another sign is a slower response to new challenges. Where once a team could solve a problem swiftly within established procedures, now every adjustment seems to require additional coordination or a special approval. Problems that would have been incidental a few years ago start to cause noticeable slowdowns. Importantly, none of these symptoms announce themselves as outright failures. Performance metrics might remain acceptable, and on the surface everything appears stable. But the absence of immediate crisis can mask a gradual decline. The organization becomes increasingly dependent on a few exceptional individuals who can paper over cracks and keep things moving. Their heroic efforts, rushing in to troubleshoot, making last-minute fixes, negotiating exceptions, maintain the illusion that the system is as safe as ever.

For example, imagine a department that continues to hit its quarterly targets, giving the impression of business-as-usual success. Beneath that surface, the manager might be scheduling extra calls each week to triage issues that never used to occur, and a handful of senior staff quietly work overtime to patch up process gaps. Nothing has "broken" in a headline-worthy way, but the margin for error is thinning. Waiting to act feels safe precisely because these changes happen slowly. Decline wears a disguise of routine. Leaders see hard-working teams eventually hitting their numbers and interpret that as proof that stability is working. In reality, the most dangerous risks are those that grow unnoticed while day-to-day performance still looks fine. Each passing quarter of "okay" results can reinforce the belief that no major change is needed. All the while, hidden fragility is building. By the time the cracks become visible, when targets start getting missed or a key player burns out, the organization may already be on the back foot, facing a risk that has matured in silence.

WHEN COORDINATION COST BECOMES A COMPETITIVE DISADVANTAGE

The internal strains of high coordination effort do not remain internal forever. Chapter 4 described how coordination costs have, over time, overtaken direct execution costs in many organizations. An organization that continues to rely on heavy alignment, layered approvals, and extensive oversight begins to lose agility in a changing environment. While the company is busy coordinating its every move, more adaptive competitors or new entrants operate with leaner decision loops. The difference is not merely speed for its own sake, it is adaptability and responsiveness. In a dynamic market or a shifting regulatory landscape, the ability to respond quickly and adjust course is a competitive advantage. Organizations encumbered by lengthy coordination find that by the time they agree on a decision or process a change through all the layers, the opportunity has passed or the problem has evolved.

Consider an established firm launching a new product feature. Under a traditional, process-heavy model, the idea may need approval from multiple committees, detailed review by every stakeholder department, and a pilot phase constrained by strict protocols. By contrast, a more nimble competitor can deploy a similar idea after a few quick iterations and a single leadership sign-off. The end result is that the second mover is first to market. It's not that the careful firm lacked insight or resources, it simply couldn't mobilize them as fast. Similarly, think about responding to an unexpected challenge like a sudden change in regulations or a supply chain disruption. A coordination-bound organization might convene task forces and steering groups to assess the situation, whereas a leaner peer starts reconfiguring operations immediately. By the time the former has aligned on a plan, the latter has already adapted.

This disadvantage of slow coordination does not announce itself overnight. It accumulates quietly. A firm with heavy processes might still deliver results for a time, but cracks show in subtle ways. Perhaps a product launch is delayed by months due to those committee reviews, giving competitors a head start. Or a customer issue that could be resolved in days stretches into weeks because it crosses multiple departmental boundaries and no one is empowered to cut through the red tape. Initially, these delays are rationalized as the "price of doing business properly," a trade-off for thoroughness and risk management. But as the external environment continues to accelerate, that price keeps rising while its benefits (like added safety or quality) start to plateau. Slowly, clients and partners notice that this organization requires more lead time for the same requests. Internally, teams grow frustrated seeing their company consistently a step behind more nimble peers in the industry, even when they have equal or greater talent and resources.

The effects of excessive coordination cost compound over time. Every extra cycle of approval or alignment is another chance for a faster-moving competitor to get ahead or for conditions to change again. What begins as a slight lag can widen into a significant gap in performance and innovation. By the time the loss of ground is undeniable, say, in declining market share or missed strategic opportunities, the disadvantage has been baked into the company's ways of working. The structures and habits that cause slowness have become deeply ingrained. Reversing them is not as simple as issuing a mandate to "move faster." It would mean untangling processes, reconsidering governance, and shifting cultural expectations that have been reinforced over years. In this way, coordination overhead quietly erodes competitiveness until the challenge of fixing it becomes formidable.

CONSENSUS THEATER BECOMES INSTITUTIONALIZED

When structural change is postponed for too long, the coping mechanisms an organization developed under strain can harden into standard practice. In Chapter 5, we introduced the idea of consensus theater, those performative routines of meetings, reviews, and approvals that give the appearance of diligent control without necessarily influencing outcomes. Initially, consensus theater emerges as a response to rising complexity and uncertainty; it is an attempt to reassure leadership that everything is under control. Over time, if no deeper changes are made, these performances stop being seen as temporary measures or necessary evils. They become simply "how we operate."

Institutionalized consensus theater means layers of oversight and ritual have become woven into the fabric of daily work. Every project update entails a gauntlet of status meetings. Every minor decision is elevated to a formal review for the sake of procedure. Leaders find their calendars filled with standing meetings that exist more to signal involvement and thoroughness than to solve problems. What began as a way to cope with strain quietly turns into a cultural norm. Employees who have spent years in this environment may not even recall a time when approvals and presentations weren't a constant presence. New hires quickly learn that success often requires mastering these rituals. They discover that preparing immaculate slide decks, following all the sign-off protocols, and engaging in extensive pre-meeting alignment is as important to their advancement as technical skill or creative insight. In effect, performing well in the theater becomes a competency of its own.

It's not that the meetings and reviews are entirely pointless, often they originated from real needs to manage risk or share information. But once they calcify into habit, their efficacy diminishes. People attend because they must, not because the forum adds value. For example, a weekly governance meeting might draw dozens of managers who each spend hours preparing reports, only for the meeting to conclude with no decisions made, just a ritual affirmation that all is "under review." Everyone involved feels they have done their duty by participating, yet the substantive work remains where it was. Over months and years, this pattern trains the organization to equate activity with progress and formality with rigor.

The danger of institutionalizing these habits is that they crowd out real decision-making and learning. With so much time and energy devoted to preparation and coordination, less bandwidth remains to grapple with the underlying issues or to experiment with new ideas. Critical decisions start migrating outside the formal channels, taken quietly by small groups who bypass the red tape out of necessity, or they get delayed indefinitely because the official process is too cumbersome to tackle them in time. In other cases, decisions simply stagnate; everyone assumes that somewhere in the maze of forums and reports, someone else will eventually take action. By normalizing the performative aspects of decision-making, the organization risks losing sight of its original purpose: making effective decisions that drive outcomes. Meanwhile, questioning these rituals becomes difficult. Proposing to cancel a standing meeting or simplify an approval flow can be viewed as heresy, as if one is undermining diligence and accountability. The theater persists in part because it feels safer to follow established routines, even if they yield diminishing returns.

Once these performative routines are ingrained, even well-intentioned leaders can mistake the volume of activity and information for actual control. When every issue has a meeting and every decision has a report, it creates a comforting impression that the organization is thoroughly managed. But this can be a false sense of security. Real strategic movement slows to a crawl, buried under self-imposed process. The organization may be busy polishing its scripts and stage sets while the world changes outside the theater. Trapped in this comfortable routine, an organization may even seek solace in adding more layers of oversight or advanced monitoring, sometimes leveraging new technology, to reinforce the feeling of control. This tendency creates the impression that if everything is documented, tracked, and regulated, then the organization must be on top of things. In reality, these additions often merely add to the illusion of control, even as true agility and influence continue to decay.

DIGITIZED CONTROL MASKS DECLINING INFLUENCE

Technology often provides a veneer of improvement, especially when an organization digitizes its existing processes without changing their underlying design. Chapter 6 explained how simply digitizing the old model can preserve its shortcomings under a modern gloss. In organizations slow to change their structures, digital dashboards, automated workflows, and elaborate reporting systems can maintain the appearance of effective control even as actual influence over outcomes wanes. Leaders may take comfort in the wealth of data at their fingertips, they can monitor every key performance indicator in real time, receive automated alerts, and enforce compliance through software. On the surface, nothing seems out of hand; if anything, the organization appears more controlled and measured than ever.

Yet this very visibility can obscure the reality that the organization's leverage is decreasing. Knowing the status of every project does not equate to being able to change those outcomes for the better. A manager might see on a dashboard that a critical initiative is behind schedule by a week, but if addressing the delay requires three levels of approval and multiple committee meetings (all now done through a modern online system), that knowledge doesn't translate into swift action. The process is as slow as it ever was, it's just meticulously documented now. Digital systems can give leaders the satisfying sense that they are "on top of things" while masking the fact that the underlying decision process remains as sluggish as before. In some cases, digitization even adds overhead: employees must now update tools, input data, and attend to the needs of the system, which creates additional micro-tasks that consume time without improving outcomes.

Consider an organization that implements an enterprise-wide dashboard to track project health. Each team updates their status weekly, flags issues, and enters risks into a database. Senior leaders receive polished reports with color-coded indicators. It feels very modern and accountable. But if a project turns red (indicating trouble), the escalation path might still be the old chain of command wrapped in new software. The team might have to fill out an electronic form to request additional resources, which then awaits approvals up the hierarchy just as a paper memo once did. The dashboard creates urgency by making the risk visible, but it doesn't cut any steps out of resolving it. In effect, everyone can see the problem on screen, yet the organization cannot react any faster than before.

Digitized control can delay the recognition of deeper decline. When problems aren't immediately visible in the metrics, leaders assume all is well. And when every issue is meticulously logged and tracked, they assume each is being managed. It's easy to believe that risks are under control because there's a spreadsheet or graph for each one. The organization might celebrate having a dashboard for every risk and a workflow for every exception. But an important question goes unasked: Are these tools actually improving our ability to achieve outcomes, or just creating the impression of rigor? In a company that has moved too slowly to adapt, technology becomes a comforting mirror, reflecting back an abundance of information and activity, but revealing little about whether the organization can still influence its fate in a meaningful way.

Leaders may only realize how much influence they've lost when, despite all the dashboards blinking green, a major objective is missed or a sudden disruption finds the organization unable to respond effectively. They look around and wonder how it happened, after all, everything was being measured and reported. It is at that point that what felt like diligent management is revealed to have been an illusion of control. The data was abundant, but the ability to act on it had withered behind the scenes. In those moments, when the carefully monitored machine fails to produce the expected outcome, the organization often finds itself falling back on the one factor outside the dashboards: the extraordinary efforts of its people. It is the talent on the ground, not the digital controls, that ultimately picks up the slack when formal mechanisms come up short.

TALENT BECOMES THE PRIMARY COMPENSATING MECHANISM

As structural misalignment deepens and formal processes struggle to keep up, organizations increasingly lean on their people to fill the gaps. In the absence of an updated model, exceptional talent becomes the glue that holds things together. As discussed in Chapter 3, mastery of the existing system is both valuable and necessary, seasoned individuals know how to navigate complex processes, handle exceptions, and get things done despite bureaucracy. That same mastery, however, can turn into a crutch masking deeper problems. Instead of the system evolving to meet new demands, the burden of adaptation falls on a few capable individuals.

In a company that has delayed change, there are often a handful of "go-to" people who make the unworkable workable. They are the managers who quietly expedite approvals by calling in personal favors, the engineers who repeatedly jury-rig solutions to bypass cumbersome protocols, and the senior specialists who can sense impending issues and head them off before anyone else even sees the warning signs. These people are highly skilled and deeply experienced, and the organization depends on them far more than it openly acknowledges. Their heroic efforts are what keep projects on track and customers satisfied when the formal process would otherwise fail to deliver.

This reliance on heroics, however, is not a sustainable strategy. It creates a brittle kind of success. The more the organization counts on individuals to compensate for structural frictions, the more it courts sudden breakdowns. If one of these key people is unavailable, whether they leave the company, take a much-needed vacation, or simply reach a breaking point, work can grind to a halt or critical errors can go unnoticed. The organization has, in effect, concentrated too much knowledge and capability in a few people, creating single points of failure in an enterprise that is supposed to run on institutional processes. Furthermore, the constant need for extraordinary effort leads to burnout and disengagement. Even the most dedicated talent cannot carry the weight of a broken model indefinitely. What was once celebrated as dedication and mastery becomes a warning sign, an indication that normal operations aren't sufficient if extraordinary measures are required so often.

Over-reliance on star performers also has subtle cultural side effects. Rank-and-file team members may come to believe that only extraordinary effort yields results, so average efforts feel futile. They see that the company survives on last-minute saves and personal interventions, which can demoralize those who follow the rules. In parallel, leadership may inadvertently reinforce the cycle by lavishly praising the heroes. It's natural to reward individuals who go above and beyond, their creativity and resilience deserve credit, but this can create an illusion that the organization is doing well because "we have great people." The narrative becomes one of individual excellence triumphing over obstacles, rather than asking why the obstacles are so persistent in the first place. That recognition, while deserved on a personal level, can blind the organization to the fact that it is surviving on exceptions.

In the end, no organization can scale or endure on the shoulders of a few exhausted heroes. Leaning on talent can paper over cracks for a while, but it cannot substitute for a system designed to run effectively without constant intervention. Depending on exceptional people as the primary safety net is, by definition, unsustainable. It not only risks operational failure when those individuals are stretched too thin, but it also delays confronting the need for structural fixes. As long as the crises are averted at the last minute, the impetus to change the system can always be deferred, until the day it can't. When that breaking point arrives, when even exceptional individuals can no longer paper over the gaps, the organization is forced to confront change under far less favorable conditions. It is at this stage that the cost of waiting becomes starkly clear: there are fewer options left on how to change, and far less room to maneuver.

THE COST OF WAITING IS PAID IN OPTIONALITY

Perhaps the most insidious cost of moving too late is the loss of flexibility. Every month or year that an organization remains in its comfortable, familiar structure, it quietly forgoes opportunities to change with less pain. Choices that existed early on gradually disappear. Processes become more entrenched; layers of management solidify; dependencies between systems and teams multiply. The organization's way of working becomes not just habit, but infrastructure. Changing anything substantial now means unwinding a tightly knotted ball of yarn. By contrast, had adjustments begun earlier, each change would have met less resistance and involved fewer entanglements.

Waiting can feel like prudence, a careful leader might think they are avoiding unnecessary disruption by holding off. But in reality, delay often only defers and amplifies the disruption. When market conditions or technological shifts eventually force a transformation, it comes at a time when the organization has the least capacity to absorb it gracefully. All the "extra" time bought through caution comes due with interest. The gap between the current state and the needed future state widens with each passing quarter of inaction. Instead of a series of smaller, controlled adjustments to evolve the model, the organization now faces the prospect of a major overhaul under pressure. And because the old structure has been reinforced for so long, even identifying what to change becomes a daunting task. People have spent entire careers following certain processes and hierarchies; questioning them isn't just a technical exercise, but a deeply cultural one that can unsettle identities and power dynamics. In fact, the longer people have operated under the old model, the more its practices and assumptions feel like second nature. Changing them now can provoke greater resistance simply because habits are deeply ingrained and tied to professional identities. In this way, delay increases not just the technical difficulty of change, but the human difficulty as well.

By waiting, leaders also give up the chance to shape change on their own terms. An organization that moves deliberately but early can choose pilot projects, safe domains, or limited-scope trials to learn and adjust gradually. It retains optionality, the ability to pivot or even pause if something doesn't work as expected. For example, a peer organization that began experimenting with a new decision-making approach or structural change a few years earlier may have encountered and resolved minor issues in low-risk settings; by the time the late adopter is forced to change, those same lessons will have to be learned under far greater pressure and scrutiny. In contrast, an organization that delays until change is no longer optional has no such luxury. It must implement untested changes broadly because there's no time for gradualism or experimentation. The margin for error disappears. For example, a new regulatory requirement or a sudden market shift might finally compel action, but with a hard deadline and high stakes that preclude a phased approach. The company finds itself having to get it "right" on the first try, because a second chance may not come.

In essence, by trying to avoid the risks of early movement, the organization accumulates a different kind of risk, the risk of inflexibility. Time becomes a constraint rather than a cushion. The loss of optionality is often only recognized in hindsight, when leaders realize that the easy, low-risk paths to adaptation that existed a few years ago are no longer available. At that point, they are left with hard choices and high-stakes implementations, precisely what they hoped to avoid by waiting. The decision to not decide, to hold off year after year, turns out to be a decision after all, one that defines the options (or lack thereof) that remain.

WHAT THIS CHAPTER ESTABLISHES

Moving too slowly to embrace structural change carries risks that are just as real as the risks of moving too quickly. In the previous chapter, we saw the pitfalls of acting on a new vision before the foundations are ready; in this chapter, we have explored the equal and opposite danger of holding onto the old model for too long. Preserving the status quo in the name of stability can allow subtle problems to compound into serious threats. A leader might be tempted to think that by not changing course, they are avoiding danger, sidestepping the unknown risks of a new model. In truth, they may be trading one set of risks for another, less visible set. In a sense, moving too late often means a slow erosion of the organization's strengths, a risk that is easy to underestimate precisely because it unfolds without a single defining jolt. The comfort of the familiar can blind an organization to the gradual accumulation of fragility, competitive disadvantage, performative habits, and over-reliance on heroics. By the time those issues demand attention, options are fewer and the cost of correction is much higher.

Recognizing the risk of moving too late does not mean swinging to the opposite extreme of reckless haste. This is not a call for immediate, impulsive overhauls. Rather, it is a call to treat delay and inertia as risks in their own right, to be weighed and managed alongside the risks of change. Neither indecision nor rushing ahead constitutes responsible leadership. True stewardship lies in navigating between over-caution and over-eagerness. In practice, effective leaders continually calibrate this balance: exercising patience where it preserves integrity or buys time to learn, and pressing forward when conditions clearly warrant action. Freezing in place is as dangerous as leaping blindly forward; the better course is a deliberate path that acknowledges both the value of stability and the necessity of evolution.

With the twin dangers of premature action and prolonged inaction now clearly delineated, the question becomes how to move forward in a balanced way. The final chapter will turn to that challenge: designing a transition that does not gamble the organization's stability, yet does not cling to it until it becomes a liability. The aim is to move with purpose but also with prudence, to preserve the core strengths that make the organization effective while adapting its structures to new realities. After all the analysis and caution of the previous chapters, the task ahead is to chart a course for change that is both diligent and decisive. In doing so, an organization can step into its future without either rushing blindly or being left irrevocably behind.

12

DESIGNING A DELIBERATE TRANSITION

The aim of a deliberate transition is not to upend everything an organization has built, but to ensure the continuity of its essential outcomes under new conditions. This is not a quest for disruption or reinvention for its own sake. It is a realignment of how work gets done, so that core missions continue to be fulfilled even as execution methods and decision-making tools evolve. Now the focus turns to moving forward responsibly. It requires recognizing a dual risk: acting too early can introduce unnecessary exposure, while waiting too long can allow fragility to quietly grow. The path between those extremes is defined not by speed or caution alone, but by a commitment to preserving what must endure while adapting to what must change.

A responsible organization approaches change with a clear priority: preserve trust, meet obligations, and maintain operational integrity throughout the transition. These are non-negotiable foundations. Any structural change, no matter how promising, must be judged by whether it continues to uphold the organization's commitments to its stakeholders. This is why the rhetoric of sudden transformation can be misleading. Grand "change programs" often chase novelty or efficiency at the expense of stability. In contrast, a deliberate transition is careful to carry forward the institution's hard-won strengths. The goal is not to throw out the rulebook, but to rewrite certain sections so that the rules still make sense in a new context. It is a process of careful redesign, not a rush to dismantle.

By framing the objective as continuity of outcomes, leaders shift the conversation from "How do we change everything?" to "What should we protect, and what should we recalibrate?". This mindset respects the fact that existing structures and processes developed for good reasons: they created reliability in a world of human limits. Those reasons do not disappear overnight. Rather than rejecting the past, a deliberate transition seeks to extend the past's success into the future. It asks how the organization's proven strengths can be maintained even as certain underlying assumptions (about execution speed, decision support, or operating constraints) begin to shift.

Treating transition as realignment also means accepting that change does not happen in one sweeping gesture. Continuity is preserved through a series of intentional, sequenced adjustments rather than a single grand overhaul. Rushing to replace everything at once, tempting as it may be in times of excitement, risks breaking the very continuity one is trying to save. Responsible change, therefore, is not a one-time revolution but a phased evolution. Each step is designed to maintain stability while moving toward a new equilibrium. In practical terms, continuity is achieved through intentional sequencing of changes, not through any wholesale replacement.

How a transition is framed and carried out will make the difference between strengthening the organization or straining it. If change is approached as an ongoing realignment of structure and purpose, it can preserve continuity at every step. But if the transition is treated as a mere technology upgrade or a quick fix, even the best intentions can stall. Understanding this distinction is the first part of designing a transition that truly balances continuity with adaptation.

TRANSITIONS FAIL WHEN THEY ARE FRAMED AS TECHNOLOGY SHIFTS

One common pitfall in organizational change is treating the transition primarily as a technology project. Often, leaders roll out a major new system or automation and expect it alone will transform how work gets done. But a transition framed around technology alone is likely to stall or disappoint, because it leaves organizational assumptions untouched. Tools are important, but by themselves they do not change who decides, how decisions are made, or who is accountable for results. Without accompanying structural shifts, a technology change often merely accelerates existing patterns thereby making the same workflows faster or more visible without altering the dynamics that caused the original friction.

Experience has shown that adopting technology without redesigning policies, decision rights, and roles tends to preserve the old model rather than replace it. Indeed, many "digitization" efforts left organizations with the same approval chains and meetings as before, only now augmented by dashboards and notifications. The outward form modernized, but the underlying behaviors and bottlenecks stayed the same. The technology itself did not fail; it was simply layered onto an unchanged structure. This happened not because the tools were flawed, but because they were applied without a corresponding change in organizational framework. An organization that simply overlays new software on its existing processes will find that it can generate more data and move information around faster, yet still struggle with slow decisions and coordination overhead. In effect, the technology will amplify whatever the organization already is. If the structure is cumbersome, a new tool may highlight that cumbersomeness even more.

The lesson is that transformation is not delivered by hardware or software alone. Real, sustained improvement in performance comes from clarifying policies, shifting decision-making to the right level, and explicitly defining intents and boundaries - the structural elements that govern how technology is used. For instance, an AI that generates analysis will still leave decisions waiting for the old approvals if authority isn't adjusted, and automating a workflow without trimming unnecessary steps simply encodes a cumbersome process in code. The technology might work perfectly, but if the rules and roles around it stay the same, the outcome will stay the same as well.

A deliberate transition therefore frames technology as an enabler, not as the instigator of change. The organization's structure, its governance, policies, and allocation of responsibilities, must evolve in parallel with any technical deployment. Only when structural change leads will new tools yield different behaviors and outcomes. In other words, structure must change before behavior reliably follows. This perspective shifts the question from "Which technology should we implement?" to "What do we need to redesign so that new technology can actually make a difference?". It also raises a practical consideration: if structural adjustments are the priority, where should one begin? Not every part of the organization can or should change at once. The most prudent approach is to start where new ways of operating can be tried safely, areas where risk is lowest and feedback is fastest, before expanding further.

START WHERE RISK IS LOWEST AND LEARNING IS FAST

Even when the need for structural change is clear, a responsible organization does not begin everywhere at once. The wiser course is to start the transition in areas where the downside is limited and the lessons will be plentiful. In practice, this means identifying parts of the operation that can serve as safe proving grounds for new ways of working. These are domains where decisions are bounded in scope, outcomes can be observed and measured quickly, and any missteps can be corrected without severe consequences. A well-contained project or a single department with a focused mandate can often pilot a new policy-driven approach or a revised decision structure with manageable risk. The intent is to choose a slice of the organization that, if the experiment falters, will not put the entire enterprise in jeopardy yet if it succeeds, it will offer credible evidence for broader change.

By starting small and safe, leaders address both extremes of the risk spectrum identified earlier. Chapter 10 cautioned against moving too early in core areas: a premature overhaul of a mission-critical process could backfire badly. Focusing initial efforts on a non-critical area avoids that trap. At the same time, Chapter 11 warned of the cost of waiting too long. Launching a pilot in a controlled domain prevents paralysis; the organization is not standing still - it is learning and adapting on a modest scale, steadily reducing uncertainty. This selective approach counters the inertia of doing nothing without courting the full dangers of a wholesale plunge. It's a middle path that turns time into an ally: the organization is gaining experience and evidence, rather than simply letting optionality decay.

Several qualities make a good starting domain. Ideally it is self-contained (changes do not ripple into other operations), has clear success criteria (so results are quickly visible), entails low regulatory exposure (mistakes won't trigger serious penalties), and has recoverable failure modes (any errors can be reversed or mitigated). These conditions create a space where experimentation is not an existential threat. Under such circumstances, people are more willing to try different approaches, and leaders can genuinely say they are learning, not gambling.

Just as important as choosing the right pilot domain is explicitly treating it as a learning initiative, not a rush to implementation. Early successes and failures in these low-risk areas provide invaluable data. They reveal where policies might be unclear, where staff need more support, or where unintended outcomes arise. This feedback loop allows the organization to refine its approach before scaling up. In effect, learning precedes confidence. By the time larger or more critical parts of the organization begin their transition, both leaders and employees should have a stronger grasp of what works, what doesn't, and why.

Starting where risk is lowest also implicitly means that the rest of the organization remains under its proven structure while the pilot runs. This is by design. It ensures continuity of normal operations and preserves what still works well, even as a small part of the system is being redesigned. In this way, selective application is not a lack of ambition but a form of safety net. It lets the old and the new coexist briefly, so that success can be built upon stable ground. This approach inevitably creates a period of coexistence between old and new methods. The organization must manage that phase deliberately, protecting what remains effective even while redesigning what needs to be improved.

PROTECT WHAT STILL WORKS WHILE REDESIGNING WHAT DOES NOT

As the organization cautiously expands its transition, it will inevitably operate in two modes at once. Some parts of the business will begin adopting new structures, perhaps guided by explicit policies and automated execution, while other parts continue to run under the traditional, process-driven model. Far from being a sign of indecision, this coexistence of old and new is a deliberate phase of a deliberate transition. The purpose is to protect everything that still works effectively, even as changes are introduced where they are needed. In practical terms, leaders must simultaneously protect what must remain intact for continuity and carefully redesign what can be improved. Responsible change means balancing these two concerns at every step.

One key to this balance is recognizing which existing mechanisms are still serving a vital protective role. Earlier in the book, we established that many processes, reviews, and supervisory layers exist for good reasons as they guard against error, maintain standards, and ensure compliance in a human-centered operating model. These mechanisms should not be discarded hastily. If a certain approval or check currently prevents serious mistakes, it must remain until a reliable alternative is in place; removing such a safeguard too soon not only introduces risk, but also breeds justifiable resistance. People inside the organization know which checks and balances have been the backstop for failure; they will understandably push back if those are stripped away without proven replacements. Early chapters emphasized that process served as a form of protection in an environment of human limitations. That protective function must be respected even as new capabilities emerge.

At the same time, effective transition requires a willingness to question structures that no longer serve their intended purpose. Some processes or rules that made sense under older constraints may prove unnecessary or overly restrictive once new methods are available. Shielding these outdated structures from scrutiny can needlessly slow down progress. For example, a manual reporting step that was needed in the past to verify unreliable data may become pure overhead once data is automatically accurate. Responsible leaders should be prepared to gradually retire or relax controls that have outlived their usefulness, but only after confirming that newer practices truly cover the same ground. This cautious unwinding ensures that adaptation does not outpace the organization's capacity to manage risk.

Maintaining both old and new operating modes simultaneously calls for clear boundaries. Confusion can quickly arise if people are unsure which rules apply in a given context. To avoid this, everyone involved needs to know exactly where the new approach is in effect and where the old rules still hold. Clarity might take the form of explicitly designated pilot teams or business units, well-defined process scopes for experiments, or temporary policies that outline how exceptions will be handled during the transition. Whatever the method, the goal is to prevent the dual structure from devolving into chaos or mixed messages. When done right, having parallel modes should feel like a structured trial, not an organizational identity crisis.

Transitional tools and scaffolds can help here. It is entirely reasonable, for instance, to run the new and old process in parallel for a period and compare outcomes, or to keep a manual override available in case the automated system encounters something unexpected. These are not signs of hesitation; they are instruments of safety and learning. A pilot project might still have a human supervisor shadowing the AI decision-making engine at first, ready to step in if needed. Such measures reassure stakeholders that safeguards remain in place. Crucially, they should be presented and understood as temporary supports that are adjustable and removable as confidence in the new system grows. By planning for these supports from the outset, leaders demonstrate that they are not simply hoping for the best; they are engineering a safe environment in which change can take root.

In managing this careful balance between continuity and change, communication is paramount. People need to hear why certain things are not changing immediately (because they still work and provide security) and why other things are being tried in a new way (because there is an opportunity to do better). When employees see that leadership is preserving the fundamentals that keep the organization safe and successful, it builds trust. When they also see that outdated pain points are being addressed methodically, it builds buy-in. Over time, as confidence in the new methods increases, the protective old mechanisms can be phased out with less controversy because their purpose has been subsumed by something better.

Even during this phase of dual operation, the organization is setting the stage for more permanent changes. The coexistence should have a deliberate end-point: once lessons are learned and new structures prove themselves, the balance shifts. Protective legacy processes are retired or transformed, and the new model extends further. To reach that point safely, leaders must be explicit about which parts of the structure are changing and which remain as-is. In other words, they must clearly articulate policy and intent for both the old and new modes of work. Only with that clarity can people navigate the transition without losing track of accountability or purpose. In practice, this means that before responsibility is redistributed to new mechanisms, leadership must clearly define the policies and intent that govern them. Put simply, the organization needs unambiguous guidance in place before it can safely hand off execution under a new model.

MAKE POLICY AND INTENT EXPLICIT BEFORE DELEGATING EXECUTION

No matter how capable a new system or method might be, an organization cannot responsibly hand off work to it without first establishing a clear framework for what is allowed, expected, and intended. In a traditional operation, much of this framework lives in the minds of experienced employees or in unwritten conventions. Managers and experts fill in the gaps, interpret unclear guidelines, and apply judgment to novel situations. But when moving to a model where execution is more automated or decisions are more distributed, these unwritten rules must be translated into explicit policies and intents. In short, doing this definition work is a prerequisite to delegation. If the organization wants to entrust certain tasks to software, or allow decisions to be made without multilayered approval, it must first articulate the boundaries and goals that will govern those tasks and decisions.

Earlier chapters introduced the shifts from process to policy, from roles to decisions, and from supervision to intent. The thread running through all of them is the need for clarity. A process-heavy approach could afford a degree of ambiguity because human supervisors were always there to catch what slipped through the cracks. For example, if a policy was vaguely worded, a manager could interpret it on the fly; if a role's responsibility was unclear, an experienced employee would improvise to get the job done. That flexibility, while useful, came at the cost of consistency and scalability. In the emerging model, where we rely on well-defined rules and autonomous execution, there is less room for ad hoc interpretation. The organization must front-load the clarity that it previously pushed down to individual judgment.

The responsibility for this clarity lies squarely with leadership and those who design the transition. They must now confront questions that were once easily sidestepped: What exactly defines an acceptable decision or outcome in each domain? What risks are absolutely off-limits, and how can those be codified as constraints? How should unusual scenarios be handled under the new model? Articulating clear answers to such questions is not a technical exercise, it is a leadership responsibility. It forces a reckoning with trade-offs and priorities that may have been implicitly managed before. For instance, a company might never have codified its tolerance for a certain risk because a seasoned manager always used discretion; now that tolerance level needs to be written into a policy or a decision rule. This work can be painstaking, but it is foundational. It amounts to writing down the organization's collective judgment so that it can be applied consistently at scale.

One common surprise for organizations in this phase is realizing how many assumptions and nuances have been carried by people rather than documented in systems or manuals. For example, teams might realize they never had a firm definition of "high priority" because people relied on personal judgment, or that many edge cases were handled by escalating to a certain leader rather than by any policy. All these hidden assumptions become glaring once work is delegated to a system. Every undefined term or unstated expectation becomes a potential point of failure when the training wheels of human oversight come off.

Without explicit policies and intent, a transition will quickly revert to old habits. Imagine implementing a new automated decision system but not clearly specifying the policy boundaries within which it can act. The first time it encounters a borderline case, uncertainty will ensue and perhaps an operator steps in "just to be safe," or a supervisor halts the process to await clarification. Before long, the organization has reintroduced manual oversight on top of the automation, nullifying the benefits. This is why ambiguity at this stage inevitably forces a reversion to familiar routines. People will instinctively compensate for gaps by leaning back on process and supervision. They are not being obstinate; they are protecting the business from ill-defined decisions.

To avoid such outcomes, the transition design must include a thorough mapping and codification of intent and policy. It is far better to invest the effort in definition now than to suffer a false start and lose credibility later. Leaders should approach this as a core responsibility: their role shifts from approving individual actions to ensuring that the rules governing those actions are sound. This is an intensive form of work, often involving cross-functional discussions and revisiting foundational principles of the business. But it pays dividends by creating a solid platform on which new capabilities can operate safely.

It's important to acknowledge that organizations often underestimate how much definition work they have deferred over the years. It is easy to assume that all the important policies have been written down and all the crucial terms well-defined, until one actually begins the audit. The reality, in many cases, is that substantial institutional knowledge resides informally in people's heads or in "the way we've always done it." Uncovering and codifying this informal knowledge can be humbling, but it is not an indictment of the past. Rather, it underscores why a deliberate transition is needed. By forcing important know-how out into the open, the organization turns the hard-won wisdom of its people into a stable foundation for the future.

Once policies are clearly articulated and intent is firmly established, the organization is prepared to delegate execution with confidence. But one piece remains critical: ensuring that as decisions and actions are redistributed, accountability moves with them. After all, clarifying the rules and goals is only half the battle. We must also clarify who owns the outcomes when those rules are followed. Organizations that neglect this piece, redesigning accountability to match the new decision-making structure, risk undermining the very trust they are trying to build.

REDESIGN ACCOUNTABILITY WITHOUT DILUTING IT

As an organization shifts how decisions are made and how work is executed, it must also shift how accountability is assigned. In the traditional model, accountability often follows hierarchy: a manager or executive is responsible for the outcomes of their team, largely because that manager controls the decisions and supervises the work. When some of those decisions become automated or are pushed closer to the front lines (within defined policy boundaries), the old patterns of accountability need to be adjusted. The goal, however, is not to dilute responsibility, it is to reattach it firmly to the new points of decision and intent. In other words, accountability must move with the decision rights. It cannot be left behind, clinging to a role that no longer actually controls what is happening.

In earlier discussions, we reframed roles in terms of decisions and elevated intent as a guiding force. Now we confront a practical question: when a decision is made by a system following a policy, or by an employee empowered under new guidelines, who is answerable for the result? The answer should be clear before such a change goes live. Responsible transition means making accountability explicit wherever authority changes. If a machine-learning model approves a transaction within preset limits, perhaps the accountability lies with the policy that set those limits and the people who defined and approved that policy. If a frontline employee is allowed to make a customer concession according to new guidelines, then that employee is accountable for using good judgment within those guidelines and the leader who created the guidelines is accountable for their clarity and adequacy. In short, when execution shifts, accountability shifts in tandem, landing where the most control (or influence) now resides.

Redesigning accountability is not about finding people to blame; it is about maintaining a line of sight from decision to consequence. In fact, done properly, the transition to a new model can make accountability clearer than before. In the legacy model, failures could be passed around with diffuse responsibility. Under a decision-centric approach, it becomes easier to pinpoint breakdowns: was it a poor decision by someone using their discretion, or a flaw in the policy constraints defined by leadership? By structuring accountability around these questions, the organization ensures that responsibility remains intact even as the mechanics of decision-making evolve. Everyone will know, in advance, how success or failure will be evaluated and who will answer for it.

Of course, shifting accountability can be sensitive. It may feel to some leaders as if they are losing control or exposing themselves to risk by letting go of direct supervision. But in reality, they are trading one form of control for another. Instead of micromanaging how every task is done, they control the framework within which tasks are done and they remain accountable for that framework. This is a different posture, but not a weaker one. A department head, for example, might no longer sign off on every expenditure if spending decisions are delegated according to a policy, but that head is certainly accountable for defining a sound spending policy and for monitoring outcomes to ensure the policy is working as intended. Their responsibility shifts from each individual transaction to the integrity of the system governing transactions.

Importantly, frontline employees or automated systems that gain new authority are not being "left on their own" in a vacuum. They are operating within bounds that leadership sets, and part of redesigning accountability is making sure those operators understand their responsibility to act within those bounds and to report or flag issues when something falls outside them. In a sense, the net of accountability becomes tighter and more transparent: there is less ambiguity about who is supposed to do what. Instead of relying on vague assumptions like "the manager would have caught that," the organization explicitly defines who monitors for exceptions and who addresses issues. This clarity protects against both unfair blame and unwarranted hand-offs. It ensures that new autonomy does not create a vacuum where everyone assumes someone else is accountable.

This clarity in accountability is crucial for building and maintaining trust during the transition. People will support changes more readily if they know that accountability is clearly assigned and that leaders are not abdicating their oversight but rather repositioning it. On the other hand, unclear accountability is the fastest way to erode trust in a new system. If something goes wrong and no one knows who is answerable or worse, if the wrong person is held to account because the new structure was never crystalized, confidence in the transition will plummet. Employees might then revert to old habits, and leaders might scramble to reassert direct control, undoing progress. To avoid this, every shift in how work gets done must be accompanied by an equally clear shift in who is responsible for outcomes.

By proactively redesigning accountability, leaders send a message that the transition does not absolve anyone of responsibility instead it simply allocates responsibility in line with the new ways of working. This reassures the organization that adopting new processes or technologies does not mean escaping oversight or consequences. It means the oversight is smarter, and consequences are tied to the right causes. When accountability is well aligned, people at all levels can embrace new tools and freedoms knowing that success and failure are fairly measured and addressed. That sense of fairness and clarity is a powerful stabilizer as the organization navigates change.

Even with clear policies, intent, and accountability, not everything will go smoothly. The transition will introduce new frictions - misunderstandings, edge-case scenarios, cultural resistance - that test the new design. These friction points are not signs of failure; they are signals and feedback. The real test of leadership in this phase is how those signals are interpreted and acted upon. Do issues prompt a return to old habits, or do they lead to adjustments and learning? Understanding friction as information is the final element in mastering a deliberate transition.

EXPECT FRICTION AND TREAT IT AS SIGNAL

No major structural change unfolds without bumps along the way. As the organization experiments with new processes, policies, and roles, it will encounter resistance, errors, and moments of confusion. These instances of friction are natural. The key is not to eliminate friction, an impossible task, but to learn from it. Every time someone reverts to an old habit, every time a team hesitates or raises a concern about the new system, there is information embedded in that reaction. It may be telling you that a policy isn't clear enough, that training was insufficient, that incentives are misaligned, or that a particular risk hasn't been fully addressed. In other words, friction is a signal. Leaders committed to a deliberate transition listen to those signals closely and respond with curiosity rather than defensiveness.

It is easy to misinterpret resistance as simply a reluctance to change or a sign of stubborn culture. While sometimes people do resist out of fear or habit, more often their pushback has an underlying rationale. Consider the experienced employee who keeps bypassing a new workflow tool to handle a client issue "the old way." This might look like non-compliance, but it could be highlighting a gap in the new workflow's design - perhaps an exception that the tool doesn't yet accommodate. Another example: a department that bombards leadership with questions after a new policy rollout might be revealing that the policy language is too vague or that front-line teams were not involved in its creation and therefore don't trust it. Each point of friction is an opportunity to refine the transition plan. By involving those raising concerns in finding solutions, leaders can often turn would-be critics into partners, because people support what they help to fix. The worst response is to simply demand compliance without investigation; the best response is to ask why the friction is occurring and address that root cause.

Not all friction is human. A glitch or unexpected system output is also a form of feedback, often pointing to a scenario the design did not anticipate. These technical issues should be handled with the same calm analysis: investigate, understand, and adjust the design rather than panicking or stubbornly pushing forward.

Crucially, how leaders handle friction sets the tone for the whole organization. If people observe that raising a concern leads to punishment or that glitches are swept under the rug, they will quickly lose faith in the transition. Problems will go unreported, small issues will fester into larger ones, and a culture of silent skepticism will take hold. Conversely, if they see leadership responding to difficulties with transparency, saying, "Thank you for flagging that; let's figure out a solution", trust deepens. Teams become more willing to experiment because they know that if something isn't right, it won't be blamed on them blindly; it will be treated as a shared problem to solve. Over time, this approach creates a virtuous cycle: early frictions yield improvements, which make the system better and more trusted, which in turn reduces future friction.

It's also important to distinguish between productive friction and signs of deeper trouble. Some resistance is helpful input, but persistent widespread refusal to adopt new practices could indicate that the change has not been sufficiently justified or that it truly impedes people's ability to get work done. In those cases, leaders might need to revisit the fundamentals: Are we explaining the "why" clearly enough? Did we perhaps choose the wrong area to start with? Are the new tools genuinely ready for prime time? Taking signal from friction doesn't mean every complaint is indulged without question, but it means every complaint is heard and evaluated on its merits.

Expecting friction also means emotionally preparing the organization to encounter setbacks without panic. Leaders can set this expectation by communicating ahead of time that there will be an adjustment period, and that discovering issues is a normal part of the process. When something does go wrong, referencing that understanding ("This is exactly why we started small, so we could find these things now") helps maintain confidence. The goal is to avoid knee-jerk reactions, whether it's abandoning the new approach entirely or plowing ahead stubbornly. A measured response views friction as feedback, fixes what needs fixing, and presses on.

The true measure of success in this phase is whether trust is maintained or even strengthened as bumps are encountered. Friction handled well can actually increase trust as people see that the system improves and that leaders stand behind their words about being thoughtful and responsible. Conversely, friction mishandled (ignored, suppressed, or overreacted to) can rapidly erode the goodwill that careful planning built. In that sense, how an organization navigates its first serious challenges under the new model often determines its trajectory going forward. Leaders who interpret friction correctly and act on it thoughtfully will find their organizations growing more confident, resilient, and united in the transition. Those who do not will find themselves slipping back into the comfortable but constraining patterns of the past.

WHAT THIS CHAPTER ESTABLISHES

A deliberate transition, as described in this chapter, is fundamentally about balance. It balances continuity and change, preservation and adaptation, caution and progress. Unlike a dramatic "transformation" program that seeks to tear out the old and install the new, this approach realigns the organization gradually and thoughtfully. It preserves core strengths, trusted processes, hard-earned expertise, and proven safeguards, even as it systematically adjusts structures and practices to fit new realities of execution and decision-making. The result is not a sudden leap into an unknown future, but a steady reorientation that carries the organization's identity and obligations forward.

Responsible change is multifaceted: start small and safe rather than with a blanket rollout; run old and new modes in parallel to protect what works while redesigning what doesn't; make policies and intent explicit before handing off work to machines or autonomous teams; shift accountability to match the new decision points; and treat friction not as failure but as feedback. None of these elements is radical on its own, but together they form a distinctive posture toward change.

Ultimately, transition is a continuing posture, not a one-time project. There is no fixed end state of "done"; rather, adapting with clarity and structure becomes an ongoing way of operating. By sequencing structural shifts responsibly, the organization keeps meeting its commitments at each step. Viewing transition as a posture acknowledges that adaptation is continuous and there will always be new conditions, technologies, and risks. The aim is to be deliberate: to move forward with design and discipline instead of either haste or inertia.

It is also important to note what this approach is not: it is not a rigid playbook or a one-size-fits-all solution. Every organization will have its own specific friction points and its own timeline dictated by its environment. The principles outlined here do not provide a checklist of actions so much as a framework for thinking. They counsel discipline, sequencing, and attentiveness to signals, but they leave room for judgment in how those apply to a particular context. In that sense, this chapter does not hand the reader a map, but rather a compass. The true path will be charted by each organization for itself, but with a steady bearing toward resilient, responsible change.

By now, the reader has traveled from understanding why the current model made sense in its time, through recognizing the pressures that necessitate change, to envisioning how to navigate that change safely. The discussion has looped back on itself in a constructive way: respect for the past and optimism for the future are joined by an ethic of careful transition in the present. With the concept of deliberate transition laid out, we come full circle. The final step is simply to pause and reflect on how these pieces fit together into a coherent story of organizational adaptation. In the epilogue, that understanding is calmly consolidated, and the book ends with clarity about what it means to carry responsibility through structural evolution.

EPILOGUE: WHAT THE READER LEARNED

Modern organizations were not built on naive assumptions or arbitrary rules. The structures most readers work within - well-defined roles, multi-layered approvals, standard operating procedures - arose for very practical reasons. They were a rational response to fundamental human constraints. People, even the most skilled, are inconsistent and limited in their capacity; fatigue sets in, judgments vary, and errors occur at scale. To compensate, companies developed processes and hierarchies as compensating mechanisms that made outcomes more predictable. By dividing work into specialized roles and enforcing formal procedures, organizations could ensure that critical tasks were performed consistently and reliably. In a world of human variability, bureaucracy was not mindless red tape, it was protection. Each approval chain, checklist, and review step reduced variance and guarded against costly mistakes. These measures created an environment where a large enterprise could deliver repeatable results, day after day, no matter how many people were involved.

This model succeeded on its own terms. It enabled decades of stability, scale, and trust in execution. The emphasis on control was never about stifling people for its own sake; it was about reliability. With clear rules and oversight, companies expanded across geographies and industries without collapsing under their own weight. Quality remained high, risks were managed, and responsibilities were clear. Many of us built our careers mastering these structures, and rightly so because the systems worked. It made consistent success possible in conditions that would otherwise have been chaotic. Far from being outdated or foolish, the traditional organizational design was coherent and effective in its time. Recognizing this fact is important. Change in this context is not a repudiation of past wisdom, but an adaptation built on that wisdom. Understanding why the old model worked, why it was rational to have layers of review and detailed processes, is precisely what allows us to evolve it responsibly.

THE ENVIRONMENT CHANGED BEFORE THE MODEL DID

While the foundational model remained steady, the context around it evolved in subtle but profound ways. Over time, execution became cheaper and faster, due largely to digitization and automation, yet the mechanisms to coordinate that execution grew heavier. The cost of coordination began to overtake the cost of the work itself. A decision that once required two conversations now needs five email threads and a compliance check. A report that took days to prepare can now be generated in minutes, but aligning everyone on which version to trust can take weeks. Leaders have sensed this paradox: individuals are more efficient, but the organization as a whole can feel slower. As one observer noted, the constraint is no longer generating answers but integrating them. In other words, organizational drag, meetings, approvals, handoffs, became the bottleneck. The design assumed a human pace and human limits, but technology changed the pace without the structure catching up. What were once reasonable coordination steps have quietly multiplied into a web of interdependencies that sap momentum.

At the same time, some forms of control expanded beyond their productive value. In the effort to ensure nothing goes wrong, organizations layered on reviews, sign-offs, and documentation requirements until a point of diminishing returns was reached. In some cases, control turned into performance. Teams learned to stage-manage compliance, checking the boxes and producing immaculate paperwork that satisfied auditors, even if those rituals had little effect on actual outcomes. This phenomenon has been called compliance theater e.g. focusing more on looking compliant than truly managing risk. It emerged not out of bad intent but from the rational desire to demonstrate control. Yet over time, the accretion of such procedures added friction without commensurate benefit. Every additional form and checkpoint provided reassurance on paper while slowing down real work. The model that once protected quality began to exhibit elements of ceremony, where ensuring everything appeared under control became almost a goal in itself.

A third subtle shift was in how technology was deployed. Digitization, rather than transforming workflows, often preserved them. Early waves of enterprise software and automation dutifully encoded existing processes exactly as they were. Companies took manual forms and made them electronic, took hierarchical approval chains and reproduced them in workflow tools. In effect, many organizations "paved the cow paths" of their operations. They embedded decades-old procedures in silicon and software without questioning the underlying assumptions. If a policy required three signatures and two meetings in the analog world, it required the same in the digital world only now with email chains and virtual sign-offs. What could have been an opportunity to streamline instead sometimes ossified the status quo. The result was that new technology often amplified the very constraints it was expected to relieve: you could process requests faster, so you simply processed more of them, with all the original checkpoints still in place.

Crucially, these changes did not announce themselves with fanfare or crisis. They arrived gradually, as creeping normality. Month by month, year by year, coordination overhead rose, control steps multiplied, and legacy assumptions lingered under shiny new interfaces. Because the organization still functioned - deliverables went out, targets were met - these structural strains were easy to overlook or dismiss as temporary headaches. There was no singular breaking point to force a reckoning. That is why recognizing the shift requires a deliberate look. It is not an admission of failure to say the environment changed; it is an act of responsibility. The conditions that made the old model so successful have shifted. The model itself, however, remained largely the same. In hindsight, as one executive observed, the paradox is not that new technology failed to deliver, it's that organizational design did not keep up. Acknowledging this mismatch is the first step. It allows us to see that the call for change is not about fixing a "broken" system, but about adapting a sound system to new realities.

CONTROL DID NOT DISAPPEAR, IT MOVED

The core argument of this book has been that control in organizations is not vanishing, it is shifting its form and focus. In the traditional model, control was embedded in process, role, and supervision: following prescribed steps, defining strict roles, and watching over work in real time. In the emerging model, control lives in policy, decision frameworks, and intent. In practice, this means that instead of enforcing every action through procedure, leaders set explicit policies that define boundaries and rules for how work should be done. Instead of relying on static job descriptions and org charts to dictate who can act, they emphasize discrete decisions: identifying which decisions matter, who should make them, and under what criteria. And instead of supervising each task, they communicate a clear intent: a shared understanding of goals and constraints that guides execution without continuous oversight. Control moves upstream, into design and planning, rather than downstream into micromanagement.

Crucially, this is not a move from human to machine, but from constraint to clarity. Humans remain responsible for setting direction, making judgments, and intervening when needed. What changes is how that responsibility is exercised day-to-day. The old model achieved control by constraining human actions (through checklists, approvals, and supervision). The new model achieves control by providing clarity, through transparent policies and objectives, that both humans and automated systems can align with. A machine might perform the repetitive steps, but a person defines the rules it follows. A junior analyst might leverage an AI agent to gather data, but a senior decision-maker has clearly delineated what decision criteria to apply. The locus of control shifts to where it is most effective: in the intent and rules set at the start, rather than in interventions at every step.

This shift is evolutionary, not anarchic. Organizations still require governance and accountability; in fact, those become even more explicit. Policies can be reviewed and audited. Decision rights can be assigned and understood by all. Intent can be discussed and refined upfront. The same objectives of the old model, consistency, compliance, quality, are still achieved, but with fewer ongoing bottlenecks. By trusting a well-crafted policy instead of a dozen sign-offs, you reduce delays without reducing diligence. By empowering people to make decisions within clear guardrails, you cut out needless escalation without losing control. By aligning teams on intent, you enable adaptability without chaos. In essence, effective control has moved to a higher level. It preserves accountability (the policies and goals are set by accountable leaders) while relieving the organization of some unnecessary coordination burden. Many forward-thinking teams are already operating this way: they have shortened decision chains and clarified who decides what, removing low-value checkpoints and simplifying collaboration. The result is that work flows faster, but not recklessly because the guidance that matters is built in from the start. This reconfiguration of control is how organizations can scale complexity without collapsing under it.

TIMING, NOT DIRECTION, IS THE REAL RISK

By this point, the direction of change should be clear. What remains uncertain and critical is timing. If the shift we have discussed is inevitable in concept, the practical question is when and how to engage with it. Here lies the delicate tension for every leader. Move too early, and you expose the organization to needless volatility. Pushing major structural change before the technology or the culture is ready can trigger backlash. You might invest in capabilities that overshoot what your operations can absorb, or unsettle employees and customers by changing proven processes prematurely. Moving too early can mean pouring resources into an initiative when the return isn't there yet, leaving the organization overextended and skeptical. On the other hand, move too late, and you risk the organization's relevance and resilience. Waiting too long can allow minor inefficiencies to harden into serious rigidity. Competitors or upstarts may seize the advantages first, setting standards and expectations that you must scramble to meet. The organization that delays adaptation can become fragile and so set in its ways that when change finally comes, it comes as a shock. As one strategic guide puts it, moving too early wastes effort on a landscape that isn't ready, while moving too late means the opportunity has passed or is fiercely contested by others. In structural terms, an overly hasty transition can cause internal breakage, whereas an overly delayed transition can leave the organization brittle or left behind.

Responsible leadership lives in the tension between these extremes. There is no simple formula for timing the evolution of an organizational model. It requires judgment, context awareness, and often a bit of humility. The point is not to avoid all risk, that is impossible, but to recognize which risk you are courting at any given moment and why. Leaders must continuously gauge the readiness of their people, the maturity of enabling technologies, and the signals from their environment. Acting with neither panic nor complacency, they choose a path of sequenced change: not a big bang overhaul, but a thoughtful series of steps that build capability and confidence over time. In practice, this means piloting new approaches in low-risk areas before scaling them, strengthening foundations (like data governance or training) before relying on them, and deliberately phasing changes so that each builds on the last. It also means being willing to pause or adapt if signals indicate that the organization is outpacing its capacity to absorb change. The hallmark of a responsible transition is deliberate sequencing. It's not the speed that defines success, but the calibration. Leaders who navigate this well do not ask "should we change now or later?" so much as they ask "in what order and at what pace should we change?" They understand that timing is a strategic variable. By treating it as such, they steer between the dangers of rushing recklessly ahead and falling perilously behind. In doing so, they fulfill their duty: protecting the organization's stability while guiding it forward.

THIS IS NOT AN IT STORY

One essential clarification: the structural shift we have discussed is not confined to technology departments or AI initiatives alone. It is a general organizational shift, touching every function where responsible work is done. In the same way that electricity or computers eventually transformed all corners of a business, far beyond the engineering teams, today's changes in execution and control permeate Finance, Legal, Operations, HR, and beyond. If you oversee compliance in a financial firm, you are as much a part of this story as the CIO is. If you manage a supply chain or run a customer service team, the ideas of moving from rigid processes to flexible policies, from strict role boundaries to agile decision-making, and from supervision to intent apply to your world. This is not about handing everything over to an "AI team" or treating it as a siloed IT project. It is about rethinking how work is structured across the board.

Evidence of this broad impact is already visible. Public reporting on surveys suggests that the adoption of AI and algorithmic tools is widespread across business functions, not just in IT. In an article on one global survey, 88% of organizations reported using AI in at least one core function, and more than half were using it in three or more functions. In other words, capabilities enabled by the assumption we started with - consistent execution by rules and automation - are being applied in marketing departments, legal departments, finance departments, and so on. Just as previous waves of change demanded cross-functional adaptation (think of how quality management or process improvement philosophies spread through every department), the shift described in this book is enterprise-wide. It must be owned by business leaders, not delegated away to specialists.

Ultimately, this is a story about responsibility, not technology. The tools and techniques may be new, but the challenge is familiar: how do we ensure our organizations deliver on their goals under changing conditions? That challenge belongs to everyone with accountability for outcomes. It requires the same clarity of purpose from a Chief Operating Officer as from a Chief Information Officer, the same openness to new structures from HR as from IT. By grounding the conversation in responsibility, who is accountable for what decision, what policy governs a process, and what intent guides a team, we keep the focus on what really matters. The message here is that modernizing how we run organizations is a shared endeavor. It calls on every function to engage, precisely because the benefits (and risks) will be felt everywhere. This is not a technical footnote to the business; it is the business. And approaching it as such ensures that we design changes that make sense in practice, on the front lines of work, not just in theory or in one department.

WHAT RESPONSIBILITY NOW REQUIRES

For anyone charged with real accountability, the implications of all the above can feel both empowering and daunting. The fundamental nature of one's responsibility is evolving. In the past, being "responsible" often meant diligently enforcing processes, monitoring compliance, and intervening to fix issues, essentially, keeping the existing machine running smoothly. Now, responsibility means redesigning parts of the machine while it runs. It requires a shift in focus from performing tasks to shaping the conditions under which tasks are performed. Concretely, responsibility now demands that leaders define intent more clearly than ever, translate that intent into explicit policies and guardrails, and hone their understanding of the critical decisions that drive outcomes. It also requires a new kind of patience and discipline: the restraint to not rush every new tool into action, and the foresight to introduce change in measured increments. In short, a responsible leader today must be part designer, part translator, and part steward of the organizational system.

This represents a maturation of the leadership role. It favors certain virtues over others. Judgment over urgency: instead of reacting to each new technological possibility with haste, the responsible leader exercises seasoned judgment about which innovations to adopt, when to do so, and how to do it safely. Design over reaction: instead of merely responding to problems as they arise, the leader proactively designs workflows and decision frameworks that prevent those problems or mitigate them gracefully.

And stewardship over control: instead of obsessing over each employee's every move, the leader guides the organization by setting direction and context, trusting teams (and intelligent systems) to operate within those parameters. This is not an easing of the leader's burden, quite the opposite. It is more demanding to articulate a clear intent than to issue a narrow order. It is more challenging to craft a good policy than to approve or deny individual requests on an ad-hoc basis. It takes more personal restraint to allow a well-prepared team to execute autonomously than to step in at every sign of trouble. But this is precisely what modern responsibility entails.

If anything, responsibility in this new era is more exacting, not less. Those in charge must still answer for outcomes, but now they must also answer for the system that produces those outcomes. They have to think not only about what needs to happen, but about how to ensure it can happen reliably without constant oversight. The reward for embracing this elevated form of responsibility is a healthier, more adaptable organization, one where people at all levels understand the why behind their actions, where decisions are made with clarity, and where change is approached as a continuous, managed journey rather than a one-time upheaval. The responsible leader becomes a source of clarity and stability amid change: someone who neither clings to old certainties nor chases the latest craze, but instead charts a deliberate course. This posture may not come with the immediate gratification of quick fixes or dramatic gestures. But it offers something deeper: a sustainable way to carry the organization forward, with eyes open and intent clear.

THE WORK CONTINUES

If there is one thing to carry forward from this discussion, it is that there is no tidy end-point to all of this. We are not heading toward a new steady state that, once reached, will neatly resolve all tensions. Rather, we are in a phase of ongoing evolution. The transition from process to policy, from roles to decisions, from supervision to intent is not a one-time program with a finish line. It is an adaptive journey that will continue as technology advances and as organizations learn. Conditions will keep changing, competitors will adjust, regulators will respond, and new tools will emerge, and what works today may need refinement tomorrow. This might sound daunting, but it is also normal. Organizational life has always been dynamic. Our predecessors dealt with continuous change (sometimes without naming it as such), and now it is simply our turn to do so more consciously.

In this sense, the work of leadership truly continues. The charge now is not to implement a static solution, but to remain engaged with change itself in a thoughtful way. There is no checklist at the end of this book, no "project plan" for the next decade. Instead, there is a perspective and a commitment. The perspective is that we can respect the past model and still move beyond it. The commitment is that we will approach this movement with clarity and care. The calm tone we have taken is deliberate: these shifts are significant, but they are not cause for panic or for complacency. They simply call for a steady hand. As a leader or a responsible practitioner, your role becomes one of ongoing stewardship through change. Whether you decide to take immediate steps or to observe a bit longer, the key is that you are now equipped to see the structural patterns at play. You can interpret friction not as failure but as information. You can weigh timing as a strategic factor. You can maintain accountability in new forms. In short, you can engage with these changes actively, not as a one-off initiative, but as part of your continuous responsibility.

And so, this book ends not with a grand proclamation but with an orientation. The reader by now should feel grounded in understanding what is happening and why. The mandate is not to rush out and upend your organization tomorrow, nor is it to put this volume on a shelf and carry on unchanged. It is to remain aware and intentional. The shift in how we coordinate and control work is already underway, and it will unfold gradually. Engaging with it thoughtfully has become part of the job for all of us who bear responsibility. This is an inevitability, but it need not be met with urgency only with steadiness. As these shifts continue to play out, you will likely encounter more examples of policy replacing procedure, of decisions outweighing org charts, and of intent guiding autonomous actions. When that happens, you will recognize them for what they are: the ongoing evolution of a model that you now understand. This journey has traced the logic from what was to what could be. The rest is not a conclusion, but a continuum. The work continues, and we carry on, composed and capable, into that continuing work.

APPENDIX: THE BOOK, SUMMARIZED

THE BOOK IN ONE SENTENCE

Modern organizations were deliberately built to compensate for human limitations, ensuring consistent and reliable outcomes; now, as new AI capabilities make it possible for certain tasks to be executed under clear rules with far less human supervision, effective control is shifting from process to policy, from roles to decisions, and from supervision to intent.

THE BOOK IN ONE PARAGRAPH

This book explains how the familiar structures of organizations, roles, processes, and layers of oversight, were rational responses to human limitations and succeeded in bringing stability and scale. Today, however, emerging AI-driven capabilities are quietly changing those underlying conditions by enabling some work to be done reliably through explicit rules and automation, reducing the need for constant human intervention. As a result, the traditional model, while still effective in many respects, is showing strain: the effort spent on coordination and control is rising even as the execution of tasks becomes easier. The book argues that organizations must evolve structurally in response: effective control will migrate away from detailed procedures and rigid job roles toward guiding policies and clear decision rights, and the nature of management will shift from continual supervision of work to the communication of intent and principles. This evolution is portrayed not as a sudden disruption, but as an inevitable realignment already underway - one that attentive, responsible leaders can approach calmly and proactively, before change is forced upon them by crisis.

TEN QUOTES FROM THE BOOK

"Modern organizations are designed around human constraints, not the nature of the work itself." - This statement highlights that companies have been structured primarily to offset human limitations (like inconsistent attention and judgment) rather than being shaped purely by the technical demands of their tasks.

"Structure was never about control for its own sake; it was about ensuring reliability at scale." - It emphasizes that hierarchies, approvals, and rules were put in place as rational mechanisms to make outcomes predictable and safe, not merely to give managers arbitrary power.

"Process is protection, not inefficiency." - This means that formal processes and checkpoints act as safeguards built from hard-earned experience, ensuring consistency and managing risk rather than merely adding red tape.

"Mastery of the current model is real, valuable, and still necessary." - It affirms that the skills and know-how people have developed to navigate the existing system remain critical and must be respected, even as new approaches emerge.

"Work now moves at the speed of alignment, not at the speed of execution." - This suggests that the pace of getting things done is most often limited by how fast people can coordinate and agree (alignment), rather than by how fast any individual can perform the technical work.

"When everyone is involved, no one is accountable." - It conveys that if responsibility for a decision or outcome is spread across too many people, clear ownership disappears and it becomes difficult to hold anyone accountable for the results.

"Visibility is often mistaken for control." - This warns that having more dashboards, status meetings, and reports can give leaders the feeling of control over outcomes, even when those measures do nothing to actually improve the work or its results.

"More decisions don't guarantee better outcomes when no one owns the whole." - It cautions that breaking every issue into numerous small approvals or committee decisions does not lead to better results, because without a single point of ownership, the overall direction and accountability can get lost.

"Digitization preserved the model it was meant to improve." - This observation notes that when organizations turned their processes into software and workflows, they mostly encoded the same old steps and assumptions in digital form-making work faster and more traceable, but without changing the fundamental way the organization operates.

"Effective control is shifting from process to policy, from roles to decisions, and from supervision to intent." - It encapsulates the book's central insight that the way organizations achieve reliable outcomes is evolving: success will depend more on clear rules and delegated decision-making guided by purpose, and less on rigid procedures, fixed job boundaries, or continuous managerial oversight.

THE TOP TEN THINGS READERS HAVE LEARNED FROM THE BOOK

A key lesson is that modern organizational structures were deliberately designed to compensate for human limits in consistency and judgment. Readers learn that roles, hierarchies, approval workflows, and oversight mechanisms emerged as coherent solutions to ensure reliable results at scale, given the inherent variability of human performance.

An insight is that the many processes and checkpoints inside organizations function as protective measures rather than pointless bureaucracy. The book makes clear that procedures like reviews, sign-offs, and detailed workflows were developed to manage risk and maintain quality, consciously trading off a bit of speed for much greater consistency and safety in outcomes.

Readers discover that mastering the existing model is still deeply valuable. The book shows that the experience and institutional knowledge required to navigate today's processes and structures is genuine competence: these hard-earned skills in managing risk, handling exceptions, and executing within constraints remain crucial, and any changes must build on this foundation rather than ignore it.

The book highlights a shifting economic reality: coordinating work has become more onerous even as executing work has become easier. It explains that as technology has made the execution of tasks faster and cheaper, the effort spent on aligning people, securing approvals, and synchronizing across teams now often dominates the work, meaning the traditional model incurs increasing cost and friction not because it "failed" but because its underlying economics have changed.

A major finding is how organizations respond to these strains by instinctively layering on more control and visibility, a phenomenon the book calls consensus theater. Readers see that as coordination gets harder, leaders often convene extra meetings, demand more reports, and add approval steps to regain a sense of control, even though this performative diligence usually creates only the appearance of progress and safety without materially improving the actual outcomes.

The book explains why past waves of digitization did not fix these structural issues. It shows that when companies translated their analog processes into digital workflows and software, they mostly preserved the same steps, roles, and decision points in a new medium, accelerating communications and transactions, but ultimately reinforcing the existing model instead of transforming it.

A key structural insight from the book is the emerging shift from process to policy in how work is controlled. Instead of relying on rigid, step-by-step procedures for every task, organizations will increasingly define clear policies and business rules as the guardrails for execution, allowing employees and automated systems to handle work more flexibly within agreed boundaries. This approach maintains consistency and safety not by micromanaging each step, but by embedding guidance in the form of policy that everyone (and every system) can follow.

The book outlines a shift from roles to decisions as the focal point of organizational design. Rather than organizing strictly around static job descriptions and traditional hierarchies, the emphasis moves to clarifying decision rights and decision-making processes. In practice, this means defining who needs to decide what, and under what conditions, so that work is structured by the logic of the decisions themselves. The result is that the right people (or algorithms) can make important decisions quickly and responsibly, regardless of their title or department, improving agility without sacrificing accountability.

A structural change the book describes is a shift from active supervision to intent. In place of managers constantly overseeing and double-checking each action, leaders will focus on communicating clear objectives and principles, that is, the intent behind the work, and then trust teams and AI-driven systems to execute within those guidelines. This change preserves accountability and direction (because leadership still sets the "why" and the boundaries), while reducing the need for continuous hands-on control over "how" every step is done. It means leadership becomes more about setting clarity of purpose and context, and less about second-by-second oversight.

Readers learned the importance of timing and restraint when pursuing change. The book makes clear that there are risks in moving too early, before the organization is ready or a new approach is proven, and risks in moving too late, after the environment has shifted and opportunities have passed. Effective leadership, therefore, requires sequencing changes thoughtfully and exercising judgment about when to act. In fact, the book treats patience and restraint as forms of competence: sometimes holding back or maintaining the current approach a bit longer is the wiser course, whereas in other cases a timely move to evolve structure is critical. The overarching lesson is that recognizing when and how fast to adapt is as crucial as recognizing what must change.

ABOUT THE AUTHOR

Andy Forbes helps organizations navigate complex digital transformations. With a background in computer science, over two decades of experience in the Salesforce ecosystem, and having co-founded multiple startups, Andy has spent his career at the intersection of technology and human workflows.

His writing explores the application of Artificial Intelligence to personal and professional challenges. In **The Intent Driven Organization** he focuses on how AI challenges traditional management hierarchies and offers a roadmap for leaders ready to move from supervision to intent.

Andy lives in rural Virginia, in a 200-year-old house that requires as much maintenance as any legacy IT system. He resides there with his wife and their two parti-colored poodles. When not writing or architecting IT solutions, he can be found riding his vintage Cannondale bicycles or playing with the latest high tech gadgets.

REFERENCES

Chapter 2: Process as Protection, Not Inefficiency

Leone, D. (2022, July 24). *Did you know the pre-flight checklist was first introduced by Boeing following the 1935 crash of the prototype B-17 (then known as the Model 299)? The Aviation Geek Club*. Retrieved from https://theaviationgeekclub.com/did-you-know-the-pre-flight-checklist-was-first-introduced-by-boeing-following-the-1935-crash-of-the-prototype-b-17-then-known-as-the-model-299/

Haynes, A. B., Weiser, T. G., Berry, W. R., Lipsitz, S. R., Breizat, A. H., Dellinger, E. P., ... Gawande, A. A. (2009). A surgical safety checklist to reduce morbidity and mortality in a global population. *New England Journal of Medicine, 360*(5), 491–499.

Chapter 4: When Coordination Costs More Than Execution

Hamel, G., & Zanini, M. (2016, March 22). *Top-Down Solutions Like Holacracy Won't Fix Bureaucracy. Harvard Business Review*. (Analysis of growth in managerial roles vs. other roles since 1983.)

Hamel, G., & Zanini, M. (2016, August 8). More of us are working in big bureaucratic organizations than ever before. *Harvard Business Review*. https://hbr.org/2016/08/more-of-us-are-working-in-big-bureaucratic-organizations-than-ever-before (Global survey report showing ~60% of working time spent on "work about work" and its consequences.)

Chapter 6: Why Digitization Wasn't the Fix

Cross, R., Rebele, R., & Grant, A. (2016). Collaborative overload. *Harvard Business Review, 94*(1), 74–79. (Study noting that at many companies employees spend up to 80% of their time on meetings, emails, and other collaborative activities, leaving little time for deep work.)

Epilogue: What the Reader Learned

Singla, A., Sukharevsky, A., Hall, B., Yee, L., & Chui, M. (2025, November 5). *The state of AI in 2025: Agents, innovation, and transformation. McKinsey & Company*. Retrieved from https://www.mckinsey.com/capabilities/quantumblack/our-insights/the-state-of-ai (Global survey finding 88% of organizations use AI in at least one core function, over half in multiple functions.)